SOCIAL GEOGRAPHY

SOCIAL GEOGRAPHY

Dr. Panna Lal

RANDOM PUBLICATIONS
NEW DELHI - 110 002 (INDIA)

Social Geography

ISBN 978-93-51117-42-1

Published in 2015 in India by

RANDOM PUBLICATIONS

4376-A/4B, Gali Murari Lal, Ansari Road
New Delhi-110 002
Phone: +9111-43580356, 23289044
E-mail: randomexports@gmail.com; sales@randompublications.com; info@randompublications.com

Reprinted 2024

Type Setting by: Friends Media, Delhi-110089
Printed at : Replika Press Pvt. Ltd.

Preface

Social geography is the branch of human geography that is most closely related to social theory in general and sociology in particular, dealing with the relation of social phenomena and its spatial components. Though the term itself has a tradition of more than 100 years, there is no consensus on its explicit content. In 1968, Anne Buttimer noted that "with some notable exceptions, social geography can be considered a field created and cultivated by a number of individual scholars rather than an academic tradition built up within particular schools". Since then, despite some calls for convergence centred on the structure and agency debate, its methodological, theoretical and topical diversity has spread even more, leading to numerours definitions of social geography and, therefore, contemporary scholars of the discipline identifying a great variety of different social geographies. However, as Benno Werlen remarked, these different perceptions are nothing else than different answers to the same two (sets of) questions, which refer to the spatial constitution of society on the one hand, and to the spatial expression of social processes on the other.

The different conceptions of social geography have also been overlapping with other sub-fields of geography and, to a lesser extent, sociology. When the term emerged within the Anglo-American tradition during the 1960s, it was basically applied as a synonym for the search for patterns in the distribution of social groups, thus being closely connected to urban geography and urban sociology. In the 1970s, the focus of debate within American human geography lay on political economic processes (though there also was a considerable number of accounts for a phenomenological perspective on social geography), while in the 1990s, geographical thought was heavily influenced by the "cultural turn". Both times, as Neil Smith noted, these approaches "claimed authority over the 'social'". In the American tradition, the concept of cultural geography has a much more distinguished history than social geography, and encompasses research areas that would be conceptualized as "social" elsewhere. In contrast, within some continental European traditions, social geography was and still is considered an approach to human geography rather than a sub-discipline, or even as identical to human geography in general. Social Geography is primarily concerned with the ways in which social relations, identities and inequalities are created. How these social creations vary over space and the role of space in their construction is the principle distinction between sociology and social geography. Whereas the former emphasizes society, geographers emphasize the spatial in social geography we are concerned with society and space. Admittedly, such a concern is central to the larger body of work we simply call human geography. The different conceptions of social geography have also been overlapping with other sub-fields of geography and, to a lesser extent, sociology. When the term emerged within the Anglo-American tradition during the 1960s, it was basically applied as a synonym for the search for patterns in the distribution of social groups, thus being closely connected to urban geography and urban sociology.

This book clearly outlines key concepts that all geographers should readily be able to explain. It does so in a highly accessible way. It is likely to be a text that students will return to throughout their degree.

I thank all members of my team who have helped in the preparation of the book. My special thanks go to "Random Publications" who have published the book.

— *Dr. Panna Lal*

Contents

Chapter 1

Introduction to Social Geography

Social geography is the branch of human geography that is most closely related to social theory in general and sociology in particular, dealing with the relation of social phenomena and its spatial components. Though the term itself has a tradition of more than 100 years, there is no consensus on its explicit content. In 1968, Anne Buttimer noted that "with some notable exceptions, (...) social geography can be considered a field created and cultivated by a number of individual scholars rather than an academic tradition built up within particular schools". Since then, despite some calls for convergence centred on the structure and agency debate, its methodological, theoretical and topical diversity has spread even more, leading to numerours definitions of social geography and, therefore, contemporary scholars of the discipline identifying a great variety of different *social geographies*. However, as Benno Werlen remarked, these different perceptions are nothing else than different answers to the same two (sets of) questions, which refer to the spatial constitution of society on the one hand, and to the spatial expression of social processes on the other.

The different conceptions of social geography have also been overlapping with other sub-fields of geography and, to a lesser extent, sociology. When the term emerged within the Anglo-American tradition during the 1960s, it was basically applied as a synonym for the search for patterns in the distribution of social groups, thus being closely connected to urban geography and urban sociology. In the 1970s, the focus of debate within American human geography lay on political economic processes (though there also was a considerable number of accounts for a phenomenological perspective on social geography), while in the 1990s, geographical thought was heavily influenced by the "cultural turn". Both times, as Neil Smith noted, these approaches "claimed authority over the 'social'". In the American tradition, the concept of cultural geography has a much more distinguished history than social geography, and encompasses research areas that would be conceptualized as "social" elsewhere. In contrast, within some continental European traditions, social geography was and still is considered an approach to human geography rather than a sub-discipline, or even as identical to human geography in general.

History

Before the Second World War

The term "social geography" (or rather "géographie sociale") originates from France, where it was used both by geographer Élisée Reclus and by sociologists of the Le Play School, perhaps independently from each other. In fact, the first proven occurrence of the term derives from a review of Reclus' *Nouvelle géographie universelle* from 1884, written by Paul de Rousiers, a member of the Le Play School. Reclus himself used the expression in several letters, the first one dating from 1895, and in his last work *L'Homme et la terre* from 1905. The first person to employ the term as part of a publication's title was Edmond Demolins, another member of the Le Play School, whose article *Géographie sociale de la France* was published in 1896 and 1897. After the death of Reclus as well as the main proponents of Le Play's ideas, and with Émile Durkheim turning away from his early concept of social morphology, Paul Vidal de la Blache, who noted that geography "is a science of places and not a science of men", remained the most influential figure of French geography. One of his students, Camille Vallaux, wrote the two-volume book *Géographie sociale*, published in 1908 and 1911. Jean Brunhes, one of Vidal's most influential disciples, included a level of (spatial) interactions among groups into his fourfold structure of human geography. Until the Second World War, no more theoretical framework for social geography was developed, though, leading to a concentration on rather descriptive rural and regional geography. However, Vidal's works were influential for the historical Annales School, who also shared the rural bias with the contemporary geographers, and Durkheim's concept of social morphology was later developed and set in connection with social geography by sociologists Marcel Mauss and Maurice Halbwachs.

The first person in the Anglo-American tradition to use the term "social geography" was George Wilson Hoke, whose paper *The Study of Social Geography* was published in 1907, yet there is no indication it had any academic impact. Le Play's work, however, was taken up in Britain by Patrick Geddes and Andrew John Herbertson. Percy M. Roxby, a former student of Herbertson, in 1930 identified social geography as one of human geography's four main branches. By contrast, the American academic geography of that time was dominated by the Berkeley School of Cultural Geography led by Carl O. Sauer, while the spatial distribution of social groups was already studied by the Chicago School of Sociology. Harlan H. Barrows, a geographer at the University of Chicago, nevertheless regarded social geography as one of the three major divisions of geography.

Another pre-war concept that combined elements of sociology and geography was the one established by Dutch sociologist Sebald Rudolf Steinmetz and his Amsterdam School of Sociography. However, it lacked a definitive subject, being a combination of geography and ethnography created as the more concrete counterpart to the rather theoretical sociology. In contrast, the Utrecht School of Social geography, which emerged in the early 1930s, sought to study the relationship between social groups and their living spaces.

Post-War Period

Continental Europe: In the German-language geography, this focus on the connection between social groups and the landscape was further developed by Hans Bobek and Wolfgang Hartke after the Second World War. For Bobek, groups of *Lebensformen* (patterns of life)—influenced by social factors—that formed the landscape, were at the centre of his social geographical analysis. In a similar approach, Hartke considered the landscape a source for indices or traces of certain social groups' behaviour. The best-known example of this perspective was the concept of *Sozialbrache* (social-fallow), i.e. the abandoning of tillage as an indicator for occupational shifts away from agriculture.

Though the French *Géographie Sociale* had been a great influence especially on Hartke's ideas, no such distinct school of thought formed within the French human geography. Nonetheless, Albert Demangeon paved the way for a number of more systematic conceptualizations of the field with his (posthumously published) notion that social groups ought to be within the centre of human geographical analysis. That task was carried out by Pierre George and Maximilien Sorre, among others. Then a Marxist, George's stance was dominated by a socio-economic rationale, but without the structuralist interpretations found in the works of some the French sociologists of the time. However, it was another French Marxist, the sociologist Henri Lefebvre, who introduced the concept of the (social) production of space. He had written on that and related topics since the 1930s, but fully expounded it in *La Production de L'Espace* as late as 1974. Sorre developed a schema of society related to the ecological idea of habitat, which was applied to an urban context by the sociologist Paul-Henry Chombart de Lauwe. For the Dutch geographer Christiaan van Paassen, the world consisted of socio-spatial entities of different scales formed by what he referred to as a "syn-ecological complex", an idea influenced by existentialism.

A more analytical ecological approach on human geography was the one developed by Edgar Kant in his native Estonia in the 1930s and later at Lund University, which he called "anthropo-ecology". His awareness of the temporal dimension of social life would lead to the formation of time geography through the works of Torsten Hägerstrand and Sven Godlund.

Ethnic Group

An ethnic group or ethnicity is a socially-defined category of people who identify with each other based on common ancestral, social, cultural or national experience. Membership of an ethnic group tends to be defined by a shared cultural heritage, ancestry, myth of origins, history, homeland, language (dialect), or even ideology, and manifests itself through symbolic systems such as religion, mythology and ritual, cuisine, dressing style, physical appearance, etc.

The largest ethnic groups in modern times comprise hundreds of millions of individuals (Han Chinese being the largest), while the smallest are limited to a few dozen individuals (numerous indigenous peoples worldwide). Larger ethnic groups may be subdivided into smaller sub-groups known variously as tribes or clans, which over time may become separate ethnic groups themselves due to endogamy and/or physical isolation from the parent group. Conversely, formerly separate ethnicities can merge to form a pan-ethnicity, and may eventually merge into one single ethnicity. Whether through division or amalgamation, the formation of a separate ethnic identity is referred to as ethnogenesis.

Depending on which source of group identity is emphasized to define membership, the following types of ethnic groups can be identified:

- ethno-racial group — emphasizing shared physical appearance based on genetic origins
- ethno-religious group — emphasizing shared affiliation with a particular religion (or denomination, or sect)

- ethno-linguistic group — emphasizing shared language (or dialect, or even use of script)
- ethno-national group — emphasizing a shared polity (cf. national identity)
- ethno-regional group — emphasizing distinct local sense of belonging stemming from geographic isolation

In many cases, such as the sense of Jewish peoplehood, more than one aspect determines membership.

Ethnic groups derived from the same historical founder population often continue to speak related languages and share a similar gene pool. By way of language shift, acculturation, adoption, and religious conversion, it is possible for some individuals or groups to leave one ethnic group and enter another (except for ethnic groups emphasizing racial purity as a key membership criterion).

Ethnicity is often used synonymously with ambiguous terms such as nation or people.

Terminology

In Early Modern English and until the mid 19th century, *ethnic* was used to mean heathen or pagan (in the sense of disparate "nations" which did not yet participate in the Christian oikumene), as the Septuagint used *ta ethne* ("the nations") to translate the Hebrew *goyim* "the nations, non-Hebrews, non-Jews". The Greek term in early antiquity (Homeric Greek) could refer to any large group, a *host* of men, a *band* of comrades as well as a *swarm* or *flock* of animals. In Classical Greek, the term took on a meaning comparable to the concept now expressed by "ethnic group", mostly translated as "nation, people"; only in Hellenistic Greek did the term tend to become further narrowed to refer to "foreign" or "barbarous" nations in particular (whence the later meaning "heathen, pagan").

In the 19th century, the term came to be used in the sense of "peculiar to a race, people or nation", in a return to the original Greek meaning. The sense of "different cultural groups", and in US English "racial, cultural or national minority group" arises in the 1930s to 1940s, serving as a replacement of the term race which had earlier taken this sense but was now becoming deprecated due to its association with ideological racism. The abstract *ethnicity* had been used for "paganism" in the 18th century, but now came to be express the meaning of an "ethnic character" (first recorded 1953). The term *ethnic group* was first recorded in 1935 and entered the Oxford English Dictionary in 1972. The term nationality depending on context may either be used synonymously with ethnicity, or synonymously with citizenship (in a sovereign state). The process that results in the emergence of an ethnicity is called ethnogenesis, a term in use in ethnological literature since about 1950.

Definitions and Conceptual History

Ethnography begins in classical antiquity; after early authors like Anaximander and Hecataeus of Miletus, Herodotus in ca. 480 BC laid the foundation of both historiography and ethnography of the ancient world. The Greeks at this time did not describe foreign nations but had also developed a concept of their own "ethnicity", which they grouped under the name of Hellenes. Herodotus (8.144.2) gave a famous account of what defined Greek (Hellenic) ethnic identity in his day, enumerating

1. shared descent
2. shared language
3. shared sanctuaries and sacrifices
4. shared customs

Whether ethnicity qualifies as a cultural universal is to some extent dependent on the exact definition used. According to "Challenges of Measuring an Ethnic World: Science, politics, and reality", "Ethnicity is a fundamental factor in human life: it is a phenomenon inherent in human experience." Many social scientists, such as anthropologists Fredrik Barth and Eric Wolf, do not consider ethnic identity to be universal. They regard ethnicity as a product of specific kinds of inter-group interactions, rather than an essential quality inherent to human groups.

According to Thomas Hylland Eriksen, the study of ethnicity was dominated by two distinct debates until recently.

- One is between "primordialism" and "instrumentalism". In the primordialist view, the participant perceives ethnic ties collectively, as an externally given, even coercive, social bond. The instrumentalist approach, on the other hand, treats ethnicity primarily as an ad-hoc element of a political strategy, used as a resource for interest groups for achieving secondary goals such as, for instance, an increase in wealth, power or status. This debate is still an important point of reference in Political science, although most scholars' approaches fall between the two poles.
- The second debate is between "constructivism" and "essentialism". Constructivists view national and ethnic identities as the

product of historical forces, often recent, even when the identities are presented as old. Essentialists view such identities as ontological categories defining social actors, and not the result of social action.

According to Eriksen, these debates have been superseded, especially in anthropology, by scholars' attempts to respond to increasingly politicised forms of self-representation by members of different ethnic groups and nations. This is in the context of debates over multiculturalism in countries, such as the United States and Canada, which have large immigrant populations from many different cultures, and post-colonialism in the Caribbean and South Asia.

Max Weber maintained that ethnic groups were *künstlich* (artificial, i.e. a social construct) because they were based on a subjective belief in shared *Gemeinschaft* (community). Secondly, this belief in shared Gemeinschaft did not create the group; the group created the belief. Third, group formation resulted from the drive to monopolise power and status. This was contrary to the prevailing naturalist belief of the time, which held that socio-cultural and behavioural differences between peoples stemmed from inherited traits and tendencies derived from common descent, then called "race".

Another influential theoretician of ethnicity was Fredrik Barth, whose "Ethnic Groups and Boundaries" from 1969 has been described as instrumental in spreading the usage of the term in social studies in the 1980s and 1990s. Barth went further than Weber in stressing the constructed nature of ethnicity. To Barth, ethnicity was perpetually negotiated and renegotiated by both external ascription and internal self-identification. Barth's view is that ethnic groups are not discontinuous cultural isolates, or logical *a prioris* to which people naturally belong. He wanted to part with anthropological notions of cultures as bounded entities, and ethnicity as primordialist bonds, replacing it with a focus on the interface between groups. "Ethnic Groups and Boundaries", therefore, is a focus on the interconnectedness of ethnic identities. Barth writes: "... categorical ethnic distinctions do not depend on an absence of mobility, contact and information, but do entail social processes of exclusion and incorporation whereby discrete categories are maintained despite changing participation and membership in the course of individual life histories."

In 1978, anthropologist Ronald Cohen claimed that the identification of "ethnic groups" in the usage of social scientists often reflected inaccurate labels more than indigenous realities:

> *... the named ethnic identities we accept, often unthinkingly, as basic givens in the literature are often arbitrarily, or even worse inaccurately, imposed.*

In this way, he pointed to the fact that identification of an ethnic group by outsiders, e.g. anthropologists, may not coincide with the self-identification of the members of that group. He also described that in the first decades of usage, the term ethnicity had often been used in lieu of older terms such as "cultural" or "tribal" when referring to smaller groups with shared cultural systems and shared heritage, but that "ethnicity" had the added value of being able to describe the commonalities between systems of group identity in both tribal and modern societies. Cohen also suggested that claims concerning "ethnic" identity (like earlier claims concerning "tribal" identity) are often colonialist practices and effects of the relations between colonized peoples and nation-states.

Social scientists have thus focused on how, when, and why different markers of ethnic identity become salient. Thus, anthropologist Joan Vincent observed that ethnic boundaries often have a mercurial character. Ronald Cohen concluded that ethnicity is "a series of nesting dichotomizations of inclusiveness and exclusiveness". He agrees with Joan Vincent's observation that (in Cohen's paraphrase) "Ethnicity ... can be narrowed or broadened in boundary terms in relation to the specific needs of political mobilization. This may be why descent is sometimes a marker of ethnicity, and sometimes not: which diacritic of ethnicity is salient depends on whether people are scaling ethnic boundaries up or down, and whether they are scaling them up or down depends generally on the political situation.

Approaches to Understanding Ethnicity

Different approaches to understanding ethnicity have been used by different social scientists when trying to understand the nature of ethnicity as a factor in human life and society. Examples of such approaches are: primordialism, essentialism, perennialism, constructivism, modernism and instrumentalism.

- "*Primordialism*", holds that ethnicity has existed at all times of human history and that modern ethnic groups have historical continuity into the far past. For them, the idea of ethnicity is closely linked to the idea of nations and is rooted in the pre-Weber understanding of humanity as being divided into primordially existing groups rooted by kinship and biological heritage.

- o "*Essentialist primordialism*" further holds that ethnicity is an *a priori* fact of human existence, that ethnicity precedes any human social interaction and that it is basically unchanged by it. This theory sees ethnic groups as natural, not just as historical. This understanding does not explain how and why nations and ethnic groups seemingly appear, disappear and often reappear through history. It also has problems dealing with the consequences of intermarriage, migration and colonization for the composition of modern day multi-ethnic societies.
- o "*Kinship primordialism*" holds that ethnic communities are extensions of kinship units, basically being derived by kinship or clan ties where the choices of cultural signs (language, religion, traditions) are made exactly to show this biological affinity. In this way, the myths of common biological ancestry that are a defining feature of ethnic communities are to be understood as representing actual biological history. A problem with this view on ethnicity is that it is more often than not the case that mythic origins of specific ethnic groups directly contradict the known biological history of an ethnic community.
- o "*Geertz's primordialism*", notably espoused by anthropologist Clifford Geertz, argues that humans in general attribute an overwhelming power to primordial human "givens" such as blood ties, language, territory, and cultural differences. In Geertz' opinion, ethnicity is not in itself primordial but humans perceive it as such because it is embedded in their experience of the world.

• "*Perennialism*", an approach that is primarily concerned with nationhood but tends to see nations and ethnic communities as basically the same phenomenon, holds that the nation, as a type of social and political organisation, is of an immemorial or "perennial" character. Smith (1999) distinguishes two variants: "continuous perennialism", which claims that particular nations have existed for very long spans of time, and "recurrent perennialism", which focuses on the emergence, dissolution and reappearance of nations as a recurring aspect of human history.

- o "*Perpetual perennialism*" holds that specific ethnic groups have existed continuously throughout history.
- o "*Situational perennialism*" holds that nations and ethnic groups emerge, change and vanish through the course of

history. This view holds that the concept of ethnicity is basically a tool used by political groups to manipulate resources such as wealth, power, territory or status in their particular groups' interests. Accordingly, ethnicity emerges when it is relevant as means of furthering emergent collective interests and changes according to political changes in the society. Examples of a perennialist interpretation of ethnicity are also found in Barth, and Seidner who see ethnicity as ever-changing boundaries between groups of people established through ongoing social negotiation and interaction.

- "*Instrumentalist perennialism*", while seeing ethnicity primarily as a versatile tool that identified different ethnics groups and limits through time, explains ethnicity as a mechanism of social stratification, meaning that ethnicity is the basis for a hierarchical arrangement of individuals. According to Donald Noel, a sociologist who developed a theory on the origin of ethnic stratification, ethnic stratification is a "system of stratification wherein some relatively fixed group membership (e.g., race, religion, or nationality) is utilised as a major criterion for assigning social positions". Ethnic stratification is one of many different types of social stratification, including stratification based on socio-economic status, race, or gender. According to Donald Noel, ethnic stratification will emerge only when specific ethnic groups are brought into contact with one another, and only when those groups are characterized by a high degree of ethnocentrism, competition, and differential power. Ethnocentrism is the tendency to look at the world primarily from the perspective of one's own culture, and to downgrade all other groups outside one's own culture. Some sociologists, such as Lawrence Bobo and Vincent Hutchings, say the origin of ethnic stratification lies in individual dispositions of ethnic prejudice, which relates to the theory of ethnocentrism. Continuing with Noel's theory, some degree of differential power must be present for the emergence of ethnic stratification. In other words, an inequality of power among ethnic groups means "they are of such unequal power that one is able to impose its will upon another". In addition to differential power, a degree of competition structured along ethnic lines is a prerequisite to ethnic stratification as well. The different ethnic groups must be competing for some common goal, such as power or influence, or a material interest, such as wealth or territory. Lawrence Bobo and Vincent Hutchings propose that competition is driven by self-interest and hostility, and results in inevitable stratification and conflict.

- "*Constructivism*" sees both primordialist and perennialist views as basically flawed, and rejects the notion of ethnicity as a basic human condition. It holds that ethnic groups are only products of human social interaction, maintained only in so far as they are maintained as valid social constructs in societies.
  - o "*Modernist constructivism*" correlates the emergence of ethnicity with the movement towards nationstates beginning in the early modern period. Proponents of this theory, such as Eric Hobsbawm, argue that ethnicity and notions of ethnic pride, such as nationalism, are purely modern inventions, appearing only in the modern period of world history. They hold that prior to this, ethnic homogeneity was not considered an ideal or necessary factor in the forging of large-scale societies.

Ethnicity is an important means by which people may identify with a larger group. Many social scientists, such as anthropologists Fredrik Barth and Eric Wolf, do not consider ethnic identity to be universal. They regard ethnicity as a product of specific kinds of inter-group interactions, rather than an essential quality inherent to human groups. Processes that result in the emergence of such identification are called ethnogenesis. Members of an ethnic group, on the whole, claim cultural continuities over time, although historians and cultural anthropologists have documented that many of the values, practices, and norms that imply continuity with the past are of relatively recent invention.

Ethnic groups differ from other social groups, such as subcultures, interest groups or social classes, because they emerge and change over historical periods (centuries) in a process known as ethnogenesis, a period of several generations of endogamy resulting in common ancestry (which is then sometimes cast in terms of a mythological narrative of a founding figure); ethnic identity is reinforced by reference to "boundary markers" - characteristics said to be unique to the group which set it apart from other groups.

Ethnicity and Nationality

In some cases, especially involving transnational migration, or colonial expansion, ethnicity is linked to nationality. Anthropologists and historians, following the modernist understanding of ethnicity as proposed by Ernest Gellner and Benedict Anderson see nations and nationalism as developing with the rise of the modern state system in the 17th century. They culminated in the rise of "nation-states" in which the presumptive boundaries of the nation coincided (or ideally coincided) with state boundaries. Thus, in the West, the notion of ethnicity, like race and nation, developed in the context of European colonial expansion, when mercantilism and capitalism were promoting global movements of populations at the same time that state boundaries were being more clearly and rigidly defined. In the 19th century, modern states generally sought legitimacy through their claim to represent "nations." Nation-states, however, invariably include populations that have been excluded from national life for one reason or another. Members of excluded groups, consequently, will either demand inclusion on the basis of equality, or seek autonomy, sometimes even to the extent of complete political separation in their own nation-state. Under these conditions—when people moved from one state to another, or one state conquered or colonized peoples beyond its national boundaries—ethnic groups were formed by people who identified with one nation, but lived in another state.

Multi-ethnic states can be the result of two opposite events, either the recent creation of state borders at variance with traditional tribal territories, or the recent immigration of ethnic minorities into a former nation state. Examples for the first case are found throughout Africa, where countries created during decolonisation inherited arbitrary colonial borders, but also in European countries such as Belgium or United Kingdom. Examples for the second case are countries such as Germany or the Netherlands, which were ethnically homogenous when they attained statehood but have received significant immigration during the second half of the 20th century. States such as the United Kingdom, France and Switzerland comprised distinct ethnic groups from their formation and have likewise experienced substantial immigration, resulting in what has been termed "multicultural" societies especially in large cities.

The states of the New World were multi-ethnic from the onset, as they were formed as colonies imposed on existing indigenous populations.

In recent decades feminist scholars (most notably Nira Yuval-Davis), have drawn attention to the fundamental ways in which women participate in the creation and reproduction of ethnic and national categories. Though these categories are usually discussed as belonging to the public, political sphere, they are upheld within the private, family sphere to a great extent. It is here that women act not just as biological reproducers but also as 'cultural carriers', transmitting knowledge and enforcing behaviours that belong to a specific collectivity. Women also often play a significant symbolic role in conceptions of nation or ethnicity, for example in the notion that 'women and children' constitute the kernel of a nation which must be defended in times of conflict, or in iconic figures such as Brittania or Marianne.

Ethnicity and Race

The distinction between race and ethnicity is considered highly problematic. Ethnicity is often assumed to be the cultural identity of a group, often based on language and tradition, while race is assumed to be a biological classification, based on DNA and bone structure. Race is a more controversial subject than ethnicity, due to its common political use. It is assumed that, based on power relations, there exist 'racialized ethnicities' and 'ethnicized races'. Ramán Grosfoguel (University of California, Berkeley) notes that 'racial/ethnic identity' is one concept and that concepts of race and ethnicity cannot be used as separate and autonomous categories.

Before Weber, race and ethnicity were often seen as two aspects of the same thing. Around 1900 and before the essentialist primordialist understanding of ethnicity was predominant, cultural differences between peoples were seen as being the result of inherited traits and tendencies. This was the time when "sciences" such as phrenology claimed to be able to correlate cultural and behavioural traits of different populations with their outward physical characteristics, such as the shape of the skull. With Weber's introduction of ethnicity as a social construct, race and ethnicity were divided from each other. A social belief in biologically well-defined races lingered on.

In 1950, the UNESCO statement, "The Race Question", signed by some of the internationally renowned scholars of the time (including Ashley Montagu, Claude Lévi-Strauss, Gunnar Myrdal, Julian Huxley, etc.), suggested that: "National, religious, geographic, linguistic and cultural groups do not necessarily coincide with racial groups: and the cultural traits of such groups have no demonstrated genetic connection with racial traits. Because serious errors of this kind are habitually committed when the term 'race' is used in popular parlance, it would be better when speaking of human races to drop the term 'race' altogether and speak of 'ethnic groups'."

In 1982 anthropologist David Craig Griffith summed up forty years of ethnographic research, arguing that racial and ethnic categories are symbolic markers for different ways that people from different parts of the world have been incorporated into a global economy:

The opposing interests that divide the working classes are further reinforced through appeals to "racial" and "ethnic" distinctions. Such appeals serve to allocate different categories of workers to rungs on the scale of labour markets, relegating stigmatized populations to the lower levels and insulating the higher echelons from competition from below. Capitalism did not create all the distinctions of ethnicity and race that function to set off categories of workers from one another. It is, nevertheless, the process of labour mobilization under capitalism that imparts to these distinctions their effective values.

According to Wolf, races were constructed and incorporated during the period of European mercantile expansion, and ethnic groups during the period of capitalist expansion.

Writing about the usage of the term "ethnic" in the ordinary language of Great Britain and the United States, in 1977 Wallman noted that

The term 'ethnic' popularly connotes '[race]' in Britain, only less precisely, and with a lighter value load. In North America, by contrast, '[race]' most commonly means colour, and 'ethnics' are the descendants of relatively recent immigrants from non-English-speaking countries. '[Ethnic]' is not a noun in Britain. In effect there are no 'ethnics'; there are only 'ethnic relations'.

In the U.S., the OMB defines the concept of race as outlined for the US Census as not "scientific or anthropological" and takes into account "social and cultural characteristics as well as ancestry", using "appropriate scientific methodologies" that are not "primarily biological or genetic in reference."

Ethno-National Conflict

Sometimes ethnic groups are subject to prejudicial attitudes and actions by the state or its constituents. In the 20th century, people began to argue that conflicts among ethnic groups or between members of an ethnic group and the state can and should be resolved in one of two ways. Some, like Jürgen Habermas and Bruce Barry, have argued that the legitimacy of modern states must be based on a notion of political rights of autonomous individual subjects. According to this view, the state should not acknowledge ethnic, national or racial identity but rather instead enforce political and legal equality of all individuals. Others, like Charles Taylor and Will Kymlicka, argue that the notion of the autonomous individual is itself a cultural construct. According to this view, states must recognise ethnic identity and develop processes through which the particular needs of ethnic groups can be accommodated within the boundaries of the nation-state.

The 19th century saw the development of the political ideology of ethnic nationalism, when the concept of race was tied to nationalism, first by German theorists including Johann Gottfried von Herder. Instances of societies focusing on ethnic ties, arguably to the exclusion of history or historical context, have resulted in the justification of nationalist goals. Two periods frequently cited as examples of this are the 19th century consolidation and expansion of the German Empire and the 20th century Nazi Germany. Each promoted the pan-ethnic idea that these governments were only acquiring lands that had always been inhabited by ethnic Germans. The history of late-comers to the nation-state model, such as those arising in the Near East and south-eastern Europe out of the dissolution of the Ottoman and Austro-Hungarian Empires, as well as those arising out of the former USSR, is marked by inter-ethnic conflicts. Such conflicts usually occur within multi-ethnic states, as opposed to between them, as in other regions of the world. Thus, the conflicts are often misleadingly labelled and characterized as civil wars when they are inter-ethnic conflicts in a multi-ethnic state.

Ethnic Groups by Continent

Africa

Ethnic groups in Africa number in the hundreds, each generally having its own language (or dialect of a language) and culture.

Many ethnic groups and nations of Africa qualify, although some groups are of a size larger than a tribal society. These mostly originate with the Sahelian kingdoms of the medieval period, such as that of the Akan, deriving from Bonoman (11th century) then the Kingdom of Ashanti (17th century).

Asia

There are an abundance of ethnic groups throughout Asia, with adaptations to the climate zones of Asia, which can be Arctic, subarctic, temperate, subtropical or tropical. The ethnic groups have adapted to mountains, deserts, grasslands, and forests. On the coasts of Asia, the ethnic groups have adopted various methods of harvest and transport. Some groups are primarily hunter-gatherers, some practice transhumance (nomadic lifestyle), others have been agrarian/rural for millennia and others becoming industrial/urban. Some groups/countries of Asia are completely urban (Hong Kong and Singapore). The colonization of Asia was largely ended in the 20th century, with national drives for independence and self-determination across the continent.

Europe

Europe has a large number of ethnic groups; Pan and Pfeil (2004) count 87 distinct "peoples of Europe", of which 33 form the majority population in at least one sovereign state, while the remaining 54 constitute ethnic minorities within every state they inhabit (although they may form local regional majorities within a sub-national entity). The total number of national minority populations in Europe is estimated at 105 million people, or 14% of 770 million Europeans.

A number of European countries, including France, and Switzerland do not collect information on the ethnicity of their resident population.

Russia has numerous recognised ethnic groups besides the 80% ethnic Russian majority. The largest group are the Tatars 3.8%. Many of the smaller groups are found in the Asian part of Russia.

Tribe

A tribe is viewed, historically or developmentally, as a social group existing before the development of, or outside of, states. Many people used the term "tribal society" to refer to societies organised largely on the basis of social, especially familial, descent groups. A customary tribe in these terms is a face-to-face community, relatively bound by kinship relations, reciprocal exchange, and strong ties to place.

"Tribe" is a contested term due to its roots in colonialism. The word has no shared referent, whether in political form, kinship relations or shared culture. Some argue that it conveys a negative connotation of a timeless unchanging past. To avoid these implications, some have chosen to use the terms "ethnic group", or nation instead.

In some places, such as North America and India, tribes are polities that have been granted legal recognition and limited autonomy by the state.

Tribes and States

Considerable debate takes place over how best to characterize tribes. This partly stems from perceived differences between pre-state tribes and contemporary tribes; some reflects more general controversy over cultural evolution and colonialism. In the popular imagination, tribes reflect a way of life that predates, and is more natural than that in modern states. Tribes also privilege primordial social ties, are clearly bounded, homogeneous, parochial, and stable. Thus, it was believed that tribes organise links between families (including clans and lineages), and provide them with a social and ideological basis for solidarity that is in some way more limited than that of an "ethnic group" or of a "nation". Anthropological and ethnohistorical research has challenged all of these notions.

Figure: *Two men from the Andamanese tribe of the Andaman Islands, India.*

Anthropologist Elman Service presented a system of classification for societies in all human cultures based on the evolution of social inequality and the role of the state. This system of classification contains four categories:

1. Gatherer-hunter bands, which are generally egalitarian.
2. Tribal societies in which there are some limited instances of social rank and prestige.
3. Stratified tribal societies led by chieftains.
4. Civilizations, with complex social hierarchies and organised, institutional governments.

In his 1975 study, *The Notion of the Tribe*, anthropologist Morton H. Fried provided numerous examples of tribes the members of which spoke different languages and practised different rituals, or that shared languages and rituals with members of other tribes. Similarly, he provided examples of tribes where people followed different political leaders, or followed the same leaders as members of other tribes. He concluded that tribes in general are characterized by fluid boundaries and heterogeneity, are not parochial, and are dynamic.

Fried, however, proposed that most contemporary tribes do not have their origin in pre-state tribes, but rather in pre-state bands. Such "secondary" tribes, he suggested, actually came about as modern products of state expansion. Bands comprise small, mobile, and fluid social formations with weak leadership, that do not generate surpluses, pay no taxes and support no standing army. Fried argued that secondary tribes develop in one of two ways. First, states could set them up as means to extend administrative and economic influence in their hinterland, where direct political control costs too much. States would encourage (or require) people on their frontiers to form more clearly bounded and centralized polities, because such polities could begin producing surpluses and taxes, and would have a leadership responsive to the needs of neighbouring states (the so-called "scheduled" tribes of the United States or of British India provide good examples of this). Second, bands could form "secondary" tribes as a means to defend themselves against state expansion. Members of bands would form more clearly bounded and centralized polities, because such polities could begin producing surpluses that could support a standing army that could fight against states, and they would have a leadership that could co-ordinate economic production and military activities.

Archaeologists continue to explore the development of pre-state tribes. Current research suggests that tribal structures constituted one type of adaptation to situations providing plentiful yet unpredictable resources. Such structures proved flexible enough to coordinate production and distribution of food in times of scarcity, without limiting or constraining people during times of surplus.

India Tribal Belt

The India's tribal belts refer to contiguous areas of settlement of tribal people of India which is to say groups or 'tribes' that remained genetically homogenous as opposed to other population groups that mixed widely within the Indian subcontinent

Northwest India

The Tribal Belt of Northwest India includes the states of Rajasthan, Gujarat, Maharashtra, and Karnataka. The tribal people of this region have origins which precede the ANI (Ancestral North Indian) and the ASI (Ancestral South India). In fact, the origins of these people are thought to stem back to the Harappan civilization of the Indus valley, the oldest traceable civilization of the Indian sub-continent which flourished between 3500BC and 2500BC.

The tribes of north-west India were once strong matrilineal societies. The changing fates and fortunes of these people has caused a gradual evolution to a more patriarchal code of living. These days the tribal societies generally follow the rule of patriliny. There are still, however, many strong cases of organised matriarchy in existence today. It is the women who organise matters such as relationships and marriages, the inheritance of land, and the distribution of wealth.

Central India

The Central India Tribal Belt stretches from Gujarat in the west up to Assam in the east across the states of Madhya Pradesh, Chhattisgarh and Jharkhand. It is among the poorest regions of the country. Over 90% of the Belt's tribal population is rural, with primitive agriculture.

Tribes in India

The term "tribe" here means a group of people that have lived at a particular place from time immemorial. Anthropologically the tribe is a system of social organisation which includes several local groups-villages, districts on lineage and normally includes a common territory, a common language and a common culture, a common name, political system, simple economy, religion and belief, primitive law and own education system. Constitutionally a tribe is he who has been mentioned in the scheduled list of Indian constitution under Article 342(i) and 342(ii).

"Tribals" are found in almost all the states of country. Currently there are between 258 and 540 scheduled tribe communities exists. The strength of these communities varies from 31 people of jarwa tribe to over 7 million gonds. Thus the gonds are big tribal community. Whereas the small communities comprising less than 1000 people include the andamanese, onge, oraon, munda, mina, khond, saora. According to recent study there are mainly 6 tribes in chhatisgarh they are gond, baiga, halba, kamar, bhunjia, korwa.

The quality of life of tribal people during pre-independence period was more deplorable and their main occupation was hunting, gathering of wood and forest products and primitive shifting cultivation. Due to destruction of forest and non availability of proper facilities, tribal people were forced to lead a poor quality of life. After independence with the adaptation of Indian constitution in 1950 special attention was given for the upliftment of the tribal people under the "article 48", it was mandatory on the part of the state government to make all efforts to improve economic, social, and educational standard of the tribal people.

Due to the welfare programmes tribal communities also made themselves conscious about their own clans upliftment. Now tribles are engaged in struggle for survival. They seek identity, autonomy equality and empowerment. They are moving out of ancestral lands to participate in all institution of state. All tribes or clans have their own unique cultures including language.

India is home to a large number of tribes with population of about 70 million. In terms of geographical distribution about 55% of tribals lived in central India, 28% in west, 12% in north-east India, 4% in South India and 1% elsewhere. These communities are actively working to preserve their rich cultures through broad institutional efforts.

Tribals constitute 8.14% of the total population of the country, numbering 84.51 million (2001 Census) and cover about 15% of the country's area. The fact that tribal people need special attention can be observed from their low social, economic and participatory indicators. Whether it is maternal and child mortality, size of agricultural holdings or access to drinking water and electricity, tribal communities lag far behind the general population. 52% of Tribal population is Below Poverty Line and what is staggering is that 54% tribals have no access to economic assets such as communication and transport.

Several alarming poverty indicators underline the importance of the need of livelihood generating activities based on locally available resources so that employment opportunities could be created. Recognising this need, the Ministry of Welfare (now Ministry of Tribal Affairs) established an organisation to take up marketing development activities for Non Timber forest produce (NTFP) on which tribal men and women spends most of their time and derive a major portion of his/her income. In 1987, the Tribal Cooperative Marketing Development Federation of India Limited (TRIFED) was set up with an aim to serve the interest of the tribal community and work for their socio-economic development by conducting its affairs in a professional, democratic and autonomous manner for undertaking marketing of local tribal products.

To achieve the aim of accelerating the economic development of tribal people by providing wider exposure to their art and crafts, TRIBES INDIA, the exclusive shops of tribal artifacts were set up all over India by TRIFED, showcasing and marketing the art and craft items produced by the tribal people. In India tribals are also called Adivasis.

Dialect

The term dialect is used in two distinct ways. One usage—the more common among linguists—refers to a variety of a language that is a characteristic of a particular group of the language's speakers. The term is applied most often to regional speech patterns, but a dialect may also be defined by other factors, such as social class. A dialect that is associated with a particular social class can be termed a sociolect, a dialect that is associated with a particular ethnic group can be termed as ethnolect, and a regional dialect may be termed a regiolect or topolect. According to this definition, any variety of a language constitutes "a dialect", including any standard varieties.

The other usage refers to a language that is socially subordinated to a regional or national standard language, often historically cognate to the standard, but not derived *from* it. In this sense, the standard language is not itself considered a dialect.

A framework was developed in 1967 by Heinz Kloss, Ausbau-, Abstand- and Dach-sprache, to describe speech communities, that while unified politically and/or culturally, include multiple dialects which though closely related genetically may be divergent to the point of inter-dialect unintelligibility.

A dialect is distinguished by its vocabulary, grammar, and pronunciation (phonology, including prosody). Where a distinction can be made only in terms of pronunciation (including prosody, or just prosody itself), the term *accent* is appropriate, not *dialect*. Other speech varieties include: standard languages, which are standardized for public performance (for example, a written standard); jargons, which are characterized by differences in lexicon (vocabulary); slang; patois; pidgins or argots.

The particular speech patterns used by an individual are termed an idiolect.

Standard and non-standard dialect A *standard dialect* (also known as a standardized dialect or "standard language") is a dialect that is supported by institutions. Such institutional support may include government recognition or designation; presentation as being the "correct" form of a language in schools; published grammars, dictionaries, and textbooks that set forth a correct spoken and written form; and an extensive formal literature that employs that dialect (prose, poetry, non-fiction, etc.). There may be multiple standard dialects associated with a single language. For example, Standard American English, Standard Canadian English, Standard Indian English, Standard Australian English, and Standard Philippine English may all be said to be standard dialects of the English language.

A nonstandard dialect, like a standard dialect, has a complete vocabulary, grammar, and syntax, but is usually not the beneficiary of institutional support. Examples of a nonstandard English dialect are Southern American English, Western Australian English and Scouse. The Dialect Test was designed by Joseph Wright to compare different English dialects with each other.

Dialect or Language

There is no universally accepted criterion for distinguishing a *language* from a *dialect*. A number of rough measures exist, sometimes leading to contradictory results. The distinction is therefore subjective and depends on the user's frame of reference.

The most common, and most purely linguistic, criterion is that of mutual intelligibility: two varieties are said to be dialects of the same language if being a speaker of one variety confers sufficient knowledge to understand and be understood by a speaker of the other; otherwise, they are said to be different languages. However, this definition becomes problematic in the case of dialect continua, in which it may be the case that dialect B is mutually intelligible with both dialect A and dialect C but dialects A and C are not mutually intelligible with each other. In this case the criterion of mutual intelligibility makes it impossible to decide whether A and C are dialects of the same language or not. Cases may also arise in which a speaker of dialect X can understand a speaker of dialect Y, but not vice versa; the mutual intelligibility criterion flounders here as well.

Another occasionally-used criterion for discriminating dialects from languages is that of linguistic authority, a more sociolinguistic notion. According to this definition, two varieties are considered dialects of the same language if (under at least some circumstances) they would defer to the same authority regarding some questions about their language. For instance, to learn the name of a new invention, or an obscure foreign species of plant, speakers of Bavarian German and East Franconian German might each consult a German dictionary or ask a German-speaking expert in the subject. By way of contrast, although Yiddish is classified by linguists as a language in the "Middle High German" group of languages, a Yiddish speaker would not consult a German dictionary to determine the word to use in such case.

By the definition most commonly used by linguists, any linguistic variety can be considered a "dialect" of *some* language—"everybody speaks a dialect". According to that interpretation, the criteria above merely serve to distinguish whether two varieties are dialects of the *same* language or dialects of *different* languages.

Terminology

The terms "language" and "dialect" are not necessarily mutually exclusive: There is nothing contradictory in the statement "the *language* of the Pennsylvania Dutch is a dialect of German".

There are various terms that linguists may use to avoid taking a position on whether the speech of a community is an independent language in its own right or a dialect of another language. Perhaps the most common is "variety"; "lect" is another. A more general term is "languoid", which does not distinguish between dialects, languages, and groups of languages, whether genealogically related or not.

Political Factors

In many societies, however, a particular dialect, often the sociolect of the elite class, comes to be identified as the "standard" or "proper" version of a language by those seeking to make a social distinction, and is contrasted with other varieties. As a result of this, in some contexts the term "dialect" refers specifically to varieties with low social status. In this secondary sense of "dialect", language varieties are often called *dialects* rather than *languages*:

- if they have no standard or codified form,
- if they are rarely or never used in writing (outside reported speech),
- if the speakers of the given language do not have a state of their own,
- if they lack prestige with respect to some other, often standardised, variety.

The status of "language" is not solely determined by linguistic criteria, but it is also the result of a historical and political development. Romansh came to be a written language, and therefore it is recognised as a language, even though it is very close to the Lombardic alpine dialects. An opposite example is the case of Chinese, whose variations such as Mandarin and Cantonese are often called dialects and not languages, despite their mutual unintelligibility.

Modern Nationalism, as developed especially since the French Revolution, has made the distinction between "language" and "dialect" an issue of great political importance. A group speaking a separate "language" is often seen as having a greater claim to being a separate "people", and thus to be more deserving of its own independent state, while a group speaking a "dialect" tends to be seen not as "a people" in its own right, but as a sub-group, part of a bigger people, which must content itself with regional autonomy. The distinction between language and dialect is thus inevitably made at least as much on a political basis as on a linguistic one, and can lead to great political controversy, or even armed conflict.

The Yiddish linguist Max Weinreich published the expression, *A shprakh iz a dialekt mit an armey un flot* in *YIVO Bleter* 25.1, 1945, p. 13. The significance of the political factors in any attempt at answering the question "what is a language?" is great enough to cast doubt on whether any strictly linguistic definition, without a socio-cultural approach, is possible. This is illustrated by the frequency with which the army-navy aphorism is cited.

Germany

In 18th and 19th century Germany, several thousand local languages of the continental west Germanic dialect continuum were reclassified as dialects of modern New High German although the vast majority of them were (and still are) mutually incomprehensible, despite the fact that they all existed long before New High German, which had at least in part been shaped as a compromise or mediative language between these local languages.

To support the intended process of nation building even further, a vague myth of some common Germanic original language developed, and German dialectology began to name dialect groups after presumed and real groups of historic tribes having existed from BC to about 600 AD, from which they were assumed to have descended. Linguistic, historic and archeological evidence for such connections is scarce, meanwhile several such ideas were proven false, yet they led to several pertaining misnomers in German dialectology. Today, all diverse West Germanic local languages under the Standard German umbrella are collectively referred to as "German dialects", (including Frisian and Low Saxon ones, that are closer to Dutch) the vast majority of German speakers still believe, they were variations of "original" or even Standard German.

The Balkans

The classification of speech varieties as dialects or languages and their relationship to other varieties of speech can thus be controversial and the verdicts inconsistent. English and Serbo-Croatian illustrate the point. English and Serbo-Croatian each have two major variants (British and American English, and Serbian and Croatian, respectively), along with numerous other varieties. For political reasons, analysing these varieties as "languages" or "dialects" yields inconsistent results: British and American English, spoken by close political and military allies, are almost universally regarded as dialects of a single language, whereas the standard languages of Serbia and Croatia, which differ from each other to a similar extent as the dialects of English, are being treated by some linguists from the region as distinct languages, largely because the two countries oscillate from being brotherly to being bitter enemies. (The Serbo-Croatian language article deals with this topic much more fully.)

Similar examples abound. Macedonian, although mutually intelligible with Bulgarian, certain dialects of Serbian and to a lesser extent the rest of the South Slavic dialect continuum is considered by Bulgarian linguists to be a Bulgarian dialect, in contrast with the contemporary international view, and the view in the Republic of Macedonia which regards it as a language in its own right. Nevertheless, before the establishment of a literary standard of Macedonian in 1944, in most sources in and out of Bulgaria before the Second World War, the southern Slavonic dialect continuum covering the area of today's Republic of Macedonia were referred to as Bulgarian dialects.

Lebanon

In Lebanon, a part of the Christian population considers "Lebanese" to be in some sense a distinct language from Arabic and not merely a dialect. During the civil war Christians often used Lebanese Arabic officially, and sporadically used the Latin script to write Lebanese, thus further distinguishing it from Arabic. All Lebanese laws are written in the standard literary form of Arabic, though parliamentary debate may be conducted in Lebanese Arabic.

North Africa

In Tunisia, Algeria, and Morocco, the spoken North African languages are sometimes considered more different from other Arabic dialects. Officially, North African and West Asian countries prefer to give preference to the Literary Arabic and conduct much of their political, cultural and religious life in it (adherence to Islam), and refrain from declaring each country's specific variety to be a separate language, because Literary Arabic is the liturgical language of Islam and the language of the Islamic sacred book, the Qur'an.

Ukraine

In the 19th Century, the Tsarist Government of Russia claimed that Ukrainian was merely a dialect of Russian and not a language in its own right. Since Soviet times, when Ukrainians were recognised as a separate nationality deserving of its own Soviet Republic, such linguistic-political claims had disappeared from circulation.

Moldova

There have been cases of a variety of speech being deliberately reclassified to serve political purposes. One example is Moldovan. In 1996, the Moldovan parliament, citing fears of "Romanian expansionism," rejected a proposal from President Mircea Snegur to change the name of the language to Romanian, and in 2003 a Moldovan–Romanian dictionary was published, purporting to show that the two countries speak different languages. Linguists of the Romanian Academy reacted by declaring that all the Moldovan words were also Romanian words; while in Moldova, the head of the Academy of Sciences of Moldova, Ion Bãrbuþã, described the dictionary as a politically motivated "absurdity".

Greater China

Unlike most languages that use alphabets to indicate the pronunciation, Chinese characters are developed from ideograms which do not always give hints to its pronunciation. While the written characters remained relatively consistent for the last two thousand years, the pronunciation and grammar in different regions has developed to an extent that the varieties of spoken language can be mutually incomprehensible. Linguists usually regard Chinese language as a family of languages rather than one language. As a series of migration to the south throughout the history, the regional languages of the south, including Xiang, Wu, Gan, Min, Yue(Cantonese), and Hakka often show traces of old Chinese or middle Chinese. From the Ming dynasty, Beijing has become the capital of China, and the Beijing spoken language has become more accepted as the common language. During the Republic of China, Mandarin Chinese was standardised as the official language, based on Beijing spoken language. Since then, other varieties of spoken language are often regarded as *fangyan* (dialects) although many linguists argue that they are different languages or topolects. Cantonese is still the most commonly used language in Hong Kong, Macau and among some overseas Chinese communities, while Min-nan has been accepted in Taiwan as an important local language alongside with Mandarin.

Historical Linguistics

Many historical linguists view any speech form as a dialect of the older medium of communication from which it developed. This point of view sees the modern Romance languages as dialects of Latin, modern Greek as a dialect of Ancient Greek, Tok Pisin as a dialect of English, and North Germanic as dialects of Old Norse. This paradigm is not entirely problem-free. It sees genetic relationships as paramount: the "dialects" of a "language" (which itself may be a "dialect" of a yet older language) may or may not be mutually intelligible. Moreover, a parent language may spawn several "dialects" which themselves subdivide any number of times, with some "branches" of the tree changing more rapidly than others.

This can give rise to the situation in which two dialects (defined according to this paradigm) with a somewhat distant genetic relationship are mutually more readily comprehensible than more closely related dialects. In one opinion, this pattern is clearly present among the modern Romance languages, with Italian and Spanish having a high degree of mutual comprehensibility, which neither language shares with French, despite some claiming that both languages are *genetically* closer to French than to each other: In fact, French-Italian and French-Spanish relative mutual incomprehensibility is due to French having undergone more rapid and more pervasive phonological change than have Spanish and Italian, not to real or imagined distance in genetic relationship. In fact, Italian and French share many more root words in common that do not even appear in Spanish.

For example, the Italian and French words for various foods, some family relationships, and body parts are very similar to each other, yet most of those words are completely different in Spanish. Italian "avere" and "essere" as auxiliaries for forming compound tenses are used similarly to French "avoir" and "être", Spanish only retains "haber" and has done away with "ser" in forming compound tenses, which are no longer used in either Spanish or Portuguese. However, when it comes to phonological structures, Italian and Spanish have undergone less change than French, with the result that some native speakers of Italian and Spanish may attain a degree of mutual comprehension that permits extensive communication.

Interlingua

One language, Interlingua, was developed so that the languages of Western civilization would act as its dialects. Drawing from such concepts as the international scientific vocabulary and Standard Average European, linguists developed a theory that the modern Western languages were actually dialects of a hidden or latent language. Researchers at the International Auxiliary Language Association extracted words and affixes that they considered to be part of Interlingua's vocabulary. In theory, speakers of the Western languages would understand written or spoken Interlingua immediately, without prior study, since their own languages were its dialects. This has often turned out to be true, especially, but not solely, for speakers of the Romance languages and educated speakers of English. Interlingua has also been found to assist in the learning of other languages. In one study, Swedish high school students learning Interlingua were able to translate passages from Spanish, Portuguese, and Italian that students of those languages found too difficult to understand. It should be noted, however, that the vocabulary of Interlingua extends beyond the Western language families.

Language

Language is the human ability to acquire and use complex systems of communication, and a language is any specific example of such a system. The scientific study of language is called linguistics.

The philosophy of language, such as whether words can represent experience, has been debated since Gorgias and Plato in Ancient Greece, with later thinkers such as Rousseau arguing that language came from emotions, while others like Kant held it came from logical thought. 20th-century philosophers such as Wittgenstein argued that philosophy is really the study of language. Major figures in linguistics include Ferdinand de Saussure and Noam Chomsky.

Estimates of the number of languages in the world vary between 5,000 and 7,000. However, any precise estimate depends on a partly arbitrary distinction between languages and dialects. Natural languages are spoken or signed, but any language can be encoded into secondary media using auditory, visual, or tactile stimuli – for example, in graphic writing, braille, or whistling. This is because human language is modality-independent. Depending on philosophical perspectives regarding the definition of language and meaning, when used as a general concept, "language" may refer to the cognitive ability to learn and use systems of complex communication, or to describe the set of rules that makes up these systems, or the set of utterances that can be produced from those rules. All languages rely on the process of semiosis to relate signs to particular meanings. Oral and sign languages contain a phonological system that governs how symbols are used to form sequences known as words or morphemes, and a syntactic system that governs how words and morphemes are combined to form phrases and utterances.

Human language has the properties of productivity, recursivity, and displacement, and relies entirely on social convention and learning. Its complex structure affords a much wider range of expressions than any known system of animal communication. Language is thought to have originated when early hominins started gradually changing their primate communication systems, acquiring the ability to form a theory of other minds and a shared intentionality. This development is sometimes thought to have coincided with an increase in brain volume, and many linguists see the structures of language as having evolved to serve specific communicative and social functions. Language is processed in many different locations in the human brain, but especially in Broca's and Wernicke's areas. Humans acquire language through social interaction in early childhood, and children generally speak fluently when they are approximately three years old. The use of language is deeply entrenched in human culture. Therefore, in addition to its strictly communicative uses, language also has many social and cultural uses, such as signifying group identity, social stratification, as well as social grooming and entertainment.

Languages evolve and diversify over time, and the history of their evolution can be reconstructed by comparing modern languages to determine which traits their ancestral languages must have had in order for the later developmental stages to occur. A group of languages that descend from a common ancestor is known as a language family. The Indo-European family is the most widely spoken and includes English, Spanish, Portuguese, Russian, and Hindi; the Sino-Tibetan family, which includes Mandarin Chinese, Cantonese, and many others; the Afro-Asiatic family, which includes Arabic, Amharic, Somali, and Hebrew; the Bantu languages, which include Swahili, Zulu, Shona, and hundreds of other languages spoken throughout Africa; and the Malayo-Polynesian languages, which include Indonesian, Malay, Tagalog, Malagasy, and hundreds of other languages spoken throughout the Pacific. Academic consensus holds that between 50% and 90% of languages spoken at the beginning of the twenty-first century will probably have become extinct by the year 2100.

Definitions

The English word "language" derives ultimately from Indo-European "tongue, speech, language" through Latin *lingua*, "language; tongue", and Old French *language*. The word is sometimes used to refer to codes, ciphers, and other kinds of artificially constructed communication systems such as formally defined computer languages used for computer programming. Unlike conventional human languages, a formal language in this sense is a system of signs for encoding and decoding information. This article specifically concerns the properties of natural human language as it is studied in the discipline of linguistics.

As an object of linguistic study, "language" has two primary meanings: an abstract concept, and a specific linguistic system, e.g. "French". The Swiss linguist Ferdinand de Saussure, who defined the modern discipline of linguistics, first explicitly formulated the distinction using the French word *langage* for language as a concept, *langue* as a specific instance of a language system, and *parole* for the concrete usage of speech in a particular language.

When speaking of language as a general concept, definitions can be used which stress different aspects of the phenomenon. These definitions also entail different approaches and understandings of language, and they inform different and often incompatible schools of linguistic theory. Debates about the nature and origin of language goes back to the ancient world. Greek philosophers such as Gorgias and Plato debated the relation between words, concepts and reality. Gorgias argued that language could represent neither the objective experience nor human experience, and that communication and truth were therefore impossible. Plato maintained that communication is possible because language represents ideas and concepts that exist independently of, and prior to, language.

During the Enlightenment and its debates about human origins, it became fashionable to speculate about the origin of language. Thinkers such as Rousseau and Herder argued that language had originated in the instinctive expression of emotions, and that it was originally closer to music and poetry than to the logical expression of rational thought. Rationalist philosophers such as Kant and Descartes held the opposite view. Around the turn of the 20th century, thinkers began to wonder about the role of language in shaping our experiences of the world – asking whether language simply reflects the objective structure of the world, or whether it creates concepts that it in turn imposes on our experience of the objective world. This led to the question of whether philosophical problems are really firstly linguistic problems. The resurgence of the view that language plays a significant role in the creation and circulation of concepts, and that the study of philosophy is essentially the study of language, is associated with what has been called the linguistic turn and philosophers such as Wittgenstein in 20th-century philosophy. These debates about language in relation to meaning and reference, cognition and consciousness remain active today.

Mental Faculty, Organ or Instinct

One definition sees language primarily as the mental faculty that allows humans to undertake linguistic behaviour: to learn languages and to produce and understand utterances. This definition stresses the universality of language to all humans, and it emphasizes the biological basis for the human capacity for language as a unique development of the human brain. Proponents of the view that the drive to language acquisition is innate in humans argue that this is supported by the fact that all cognitively normal children raised in an environment where language is accessible will acquire language without formal instruction. Languages may even develop spontaneously in environments where people live or grow up together without a common language; for example, creole languages and spontaneously developed sign languages such as Nicaraguan Sign Language. This view, which can be traced back to the philosophers Kant and Descartes, understands language to be largely innate, for example, in Chomsky's theory of Universal Grammar, or American philosopher Jerry Fodor's extreme innatist theory. These kinds of definitions are often applied in studies of language within a cognitive science framework and in neurolinguistics.

Formal Symbolic System

Another definition sees language as a formal system of signs governed by grammatical rules of combination to communicate meaning. This definition stresses that human languages can be described as closed structural systems consisting of rules that relate particular signs to particular meanings. This structuralist view of language was first introduced by Ferdinand de Saussure, and his structuralism remains foundational for many approaches to language.

Some proponents of Saussure's view of language have advocated a formal approach which studies language structure by identifying its basic elements and then by presenting a formal account of the rules according to which the elements combine in order to form words and sentences. The main proponent of such a theory is Noam Chomsky, the originator of the generative theory of grammar, who has defined language as the construction of sentences that can be generated using transformational grammars. Chomsky considers these rules to be an innate feature of the human mind and to constitute the rudiments of what language is. By way of contrast, such transformational grammars are also commonly used to provide formal definitions of language are commonly used in formal logic, in formal theories of grammar, and in applied computational linguistics. In the philosophy of language, the view of linguistic meaning as residing in the logical relations between propositions and reality was developed by philosophers such as Alfred Tarski, Bertrand Russell, and other formal logicians.

Tool for Communication

Yet another definition sees language as a system of communication that enables humans to exchange verbal or symbolic utterances. This definition stresses the social functions of language and the fact that humans use it to express themselves and to manipulate objects in their environment. Functional theories of grammar explain grammatical structures by their communicative functions, and understand the grammatical structures of language to be the result of an adaptive process by which grammar was "tailored" to serve the communicative needs of its users.

This view of language is associated with the study of language in pragmatic, cognitive, and interactive frameworks, as well as in sociolinguistics and linguistic anthropology. Functionalist theories tend to study grammar as dynamic phenomena, as structures that are always in the process of changing as they are employed by their speakers. This view places importance on the study of linguistic typology, or the classification of languages according to structural features, as it can be shown that processes of grammaticalization tend to follow trajectories that are partly dependent on typology. In the philosophy of language, the view of pragmatics as being central to language and meaning is often associated with Wittgenstein's later works and with ordinary language philosophers such as J. L. Austin, Paul Grice, John Searle, and W. O. Quine.

Unique Status of Human Language

Human language is unique in comparison to other forms of communication, such as those used by non-human animals. Communication systems used by other animals such as bees or apes are closed systems that consist of a finite, usually very limited, number of possible ideas that can be expressed.

In contrast, human language is open-ended and productive, meaning that it allows humans to produce a vast range of utterances from a finite set of elements, and to create new words and sentences. This is possible because human language is based on a dual code, in which a finite number of elements which are meaningless in themselves (e.g. sounds, letters or gestures) can be combined to form a theoretically infinite number of larger units of meaning (words and sentences). Furthermore, the symbols and grammatical rules of any particular language are largely arbitrary, so that the system can only be acquired through social interaction. The known systems of communication used by animals, on the other hand, can only express a finite number of utterances that are mostly genetically determined.

Several species of animals have proved to be able to acquire forms of communication through social learning: for instance a bonobo named Kanzi learned to express itself using a set of symbolic lexigrams. Similarly, many species of birds and whales learn their songs by imitating other members of their species. However, while some animals may acquire large numbers of words and symbols, none have been able to learn as many different signs as are generally known by an average 4 year old human, nor have any acquired anything resembling the complex grammar of human language.

Human languages also differ from animal communication systems in that they employ grammatical and semantic categories, such as noun and verb, present and past, which may be used to express exceedingly complex meanings. Human language is also unique in having the property of recursivity: for example, a noun phrase can contain another noun phrase or a clause can contain another clause. Human language is also the only known natural communication system whose adaptability may be referred to as *modality independent*. This means that it can be used not only for communication through one channel or medium, but through several. For example, spoken language uses the auditive modality, whereas sign languages and writing use the visual modality, and braille writing uses the tactile modality.

Human language is also unique in being able to refer to abstract concepts and to imagined or hypothetical events as well as events that took place in the past or may happen in the future. This ability to refer to events that are not at the same time or place as the speech event is called *displacement*, and while some animal communication systems can use displacement (such as the communication of bees that can communicate the location of sources of nectar that are out of sight), the degree to which it is used in human language is also considered unique.

The Study of Language

he study of language, linguistics, has been developing into a science since the first grammatical descriptions of particular languages in India more than 2000 years ago. Modern linguistics is a science that concerns itself with all aspects of language, examining it from all of the theoretical viewpoints described above.

Subdisciplines

The academic study of language is conducted within many different disciplinary areas and from different theoretical angles, all of which inform modern approaches to linguistics. For example, descriptive linguistics examines the grammar of single languages, theoretical linguistics develops theories on how best to conceptualize and define the nature of language based on data from the various extant human languages, sociolinguistics studies how languages are used for social purposes informing in turn the study of the social functions of language and grammatical description, neurolinguistics studies how language is processed in the human brain and allows the experimental testing of theories, computational linguistics builds on theoretical and descriptive linguistics to construct computational models of language often aimed at processing natural language or at testing linguistic hypotheses, and historical linguistics relies on grammatical and lexical descriptions of languages to trace their individual histories and reconstruct trees of language families by using the comparative method.

Early History

The formal study of language is often considered to have started in India with Panini, the 5th century BC grammarian who formulated 3,959 rules of Sanskrit morphology. However, Sumerian scribes already studied the differences between Sumerian and Akkadian grammar around 1900 BC. Subsequent grammatical traditions developed in all of the ancient cultures that adopted writing.

In the 17th century AD, the French Port-Royal Grammarians developed the idea that the grammars of all languages were a reflection of the universal basics of thought, and therefore that grammar was universal. In the 18th century, the first use of the comparative method by British philologist and expert on ancient India William Jones sparked the rise of comparative linguistics. The scientific study of language was broadened from Indo-European to language in general by Wilhelm von Humboldt. Early in the 20th century, Ferdinand de Saussure introduced the idea of language as a static system of interconnected units, defined through the oppositions between them.

By introducing a distinction between diachronic and synchronic analyses of language, he laid the foundation of the modern discipline of linguistics. Saussure also introduced several basic dimensions of linguistic analysis that are still fundamental in many contemporary linguistic theories, such as the distinctions between syntagm and paradigm, and the Langue-parole distinction, distinguishing language as an abstract system (*langue*), from language as a concrete manifestation of this system (*parole*).

Contemporary Linguistics

In the 1960s, Noam Chomsky formulated the generative theory of language. According to this theory, the most basic form of language is a set of syntactic rules that is universal for all humans and which underlies the grammars of all human languages. This set of rules is called Universal Grammar; for Chomsky, describing it is the primary objective of the discipline of linguistics. Thus, he considered that the grammars of individual languages are only of importance to linguistics insofar as they allow us to deduce the universal underlying rules from which the observable linguistic variability is generated.

In opposition to the formal theories of the generative school, functional theories of language propose that since language is fundamentally a tool, its structures are best analyzed and understood by reference to their functions. Formal theories of grammar seek to define the different elements of language and describe the way they relate to each other as systems of formal rules or operations, while functional theories seek to define the functions performed by language and then relate them to the linguistic elements that carry them out. The framework of cognitive linguistics interprets language in terms of the concepts (which are sometimes universal, and sometimes specific to a particular language) which underlie its forms. Cognitive linguistics is primarily concerned with how the mind creates meaning through language.

Physiological and Neural Architecture of Language and Speech

Speaking is the default modality for language in all cultures. The production of spoken language depends on sophisticated capacities for controlling the lips, tongue and other components of the vocal apparatus, the ability to acoustically decode speech sounds, and the neurological apparatus required for acquiring and producing language. The study of the genetic bases for human language is at an early stage: the only gene that has definitely been implicated in language production is FOXP2, which may cause a kind of congenital language disorder if affected by mutations.

Social Contexts of Use and Transmission

While humans have the ability to learn any language, they only do so if they grow up in an environment in which language exists and is used by others. Language is therefore dependent on communities of speakers in which children learn language from their elders and peers and themselves transmit language to their own children. Languages are used by those who speak them to communicate and to solve a plethora of social tasks. Many aspects of language use can be seen to be adapted specifically to these purposes. Due to the way in which language is transmitted between generations and within communities, language perpetually changes, diversifying into new languages or converging due to language contact. The process is similar to the process of evolution, where the process of descent with modification leads to the formation of a phylogenetic tree.

However, languages differ from a biological organisms in that they readily incorporate elements from other languages through the process of diffusion, as speakers of different languages come into contact. Humans also frequently speak more than one language, acquiring their first language or languages as children, or learning new languages as they grow up. Because of the increased language contact in the globalizing world, many small languages are becoming endangered as their speakers shift to other languages that afford the possibility to participate in larger and more influential speech communities.

Usage and Meaning

The semantic study of meaning assumes that meaning is located in a relation between signs and meanings that are firmly established through social convention. However, semantics does not study the way in which social conventions are made and affect language. Rather, when studying the way in which words and signs are used, it is often the case that words have different meanings, depending on the social context of use. An important example of this is the process called deixis, which describes the way in which certain words refer to entities through their relation between a specific point in time and space when the word is uttered. Such words are, for example, the word, "I" (which designates the person speaking), "now" (which designates the moment of speaking), and "here" (which designates the time of speaking). Signs also change their meanings over time, as the conventions governing their usage gradually change. The study of how the meaning of linguistic expressions changes depending on context is called pragmatics. Deixis is an important part of the way that we use language to point out entities in the world. Pragmatics is concerned with the ways in which language use is patterned and how these patterns contribute to meaning. For example, in all languages, linguistic expressions can be used not just to transmit information, but to perform actions. Certain actions are made only through language, but nonetheless have tangible effects, e.g. the act of "naming", which creates a new name for some entity, or the act of "pronouncing someone man and wife", which creates a social contract of marriage. These types of acts are called speech acts, although they can of course also be carried out through writing or hand signing.

The form of linguistic expression often does not correspond to the meaning that it actually has in a social context. For example, if at a dinner table a person asks, "Can you reach the salt?", that is, in fact, not a question about the length of the arms of the one being addressed, but a request to pass the salt across the table. This meaning is implied by the context in which it is spoken; these kinds of effects of meaning are called conversational implicatures. These social rules for which ways of using language are considered appropriate in certain situations and how utterances are to be understood in relation to their context vary between communities, and learning them is a large part of acquiring communicative competence in a language.

Language Acquisition

All healthy, normally developing human beings learn to use language. Children acquire the language or languages used around them: whichever languages they receive sufficient exposure to during childhood. The development is essentially the same for children acquiring sign or oral languages. This learning process is referred to as first-language acquisition, since unlike many other kinds of learning, it requires no direct teaching or specialised study. In *The Descent of Man*, naturalist Charles Darwin called this process "an instinctive tendency to acquire an art".

First language acquisition proceeds in a fairly regular sequence, though there is a wide degree of variation in the timing of particular stages among normally developing infants. From birth, newborns respond more readily to human speech than to other sounds. Around one month of age, babies appear to be able to distinguish between different speech sounds. Around six months of age, a child will begin babbling, producing the speech sounds or handshapes of the languages used around them. Words appear around the age of 12 to 18 months; the average vocabulary of an eighteen-month old child is around 50 words. A child's first utterances are holophrases (literally "whole-sentences"), utterances that use just one word to communicate some idea. Several months after a child begins producing words, she or he will produce two-word utterances, and within a few more months will begin to produce telegraphic speech, or short sentences that are less grammatically complex than adult speech, but that do show regular syntactic structure. From roughly the age of three to five years, a child's ability to speak or sign is refined to the point that it resembles adult language.

Acquisition of second and additional languages can come at any age, through exposure in daily life or courses. Children learning a second language are more likely to achieve native-like fluency than adults, but in general, it is very rare for someone speaking a second language to pass completely for a native speaker. An important difference between first language acquisition and additional language acquisition is that the process of additional language acquisition is influenced by languages that the learner already knows.

Language and Culture

Languages, understood as the particular set of speech norms of a particular community, are also a part of the larger culture of the community that speaks them. Languages do not differ only in pronunciation, vocabulary, or grammar, but also through having different "cultures of speaking". Humans use language as a way of signalling identity with one cultural group and difference from others. Even among speakers of one language, several different ways of using the language exist, and each is used to signal affiliation with particular subgroups within a larger culture. Linguists and anthropologists, particularly sociolinguists, ethnolinguists, and linguistic anthropologists have specialised in studying how ways of speaking vary between speech communities.

Linguists use the term "varieties" to refer to the different ways of speaking a language. This term includes geographically or socioculturally defined dialects as well as the jargons or styles of subcultures. Linguistic anthropologists and sociologists of language define communicative style as the ways that language is used and understood within a particular culture.

Because norms for language use are shared by members of a specific group, communicative style also becomes a way of displaying and constructing group identity. Linguistic differences may become salient markers of divisions between social groups, for example, speaking a language with a particular accent may imply membership of an ethnic minority or social class, one's area of origin, or status as a second language speaker. These kinds of differences are not part of the linguistic system, but are an important part of how people use language as a social tool for constructing groups.

However, many languages also have grammatical conventions that signal the social position of the speaker in relation to others through the use of registers that are related to social hierarchies or divisions. In many languages, there are stylistic or even grammatical differences between the ways men and women speak, between age groups, or between social classes, just as some languages employ different words depending on who is listening. For example, in the Australian language Dyirbal, a married man must use a special set of words to refer to everyday items when speaking in the presence of his mother-in-law. Some cultures, for example, have elaborate systems of "social deixis", or systems of signalling social distance through linguistic means. In English, social deixis is shown mostly through distinguishing between addressing some people by first name and others by surname, and in titles such as "Mrs.", "boy", "Doctor", or "Your Honour", but in other languages, such systems may be highly complex and codified in the entire grammar and vocabulary of the language. For instance, in languages of east Asia such as Thai, Burmese, and Javanese, different words are used according to whether a speaker is addressing someone of higher or lower rank than oneself in a ranking system with animals and children ranking the lowest and gods and members of royalty as the highest.

Language Contact

One important source of language change is contact and resulting diffusion of linguistic traits between languages. Language contact occurs when speakers of two or more languages or varieties interact on a regular basis. Multilingualism is likely to have been the norm throughout human history and most people in the modern world are multilingual. Before the rise of the concept of the ethno-national state, monolingualism was characteristic mainly of populations inhabiting small islands. But with the ideology that made one people, one state, and one language the most desirable political arrangement, monolingualism started to spread throughout the world. Nonetheless, there are only 250 countries in the world corresponding to some 6000 languages, which means that most countries are multilingual and most languages therefore exist in close contact with other languages.

When speakers of different languages interact closely, it is typical for their languages to influence each other. Through sustained language contact over long periods, linguistic traits diffuse between languages, and languages belonging to different families may converge to become more similar. In areas where many languages are in close contact, this may lead to the formation of language areas in which unrelated languages share a number of linguistic features. A number of such language areas have been documented, among them, the Balkan language area, the Mesoamerican language area, and the Ethiopian language area. Also, larger areas such as South Asia, Europe, and Southeast Asia have sometimes been considered language areas, because of widespread diffusion of specific areal features.

Language contact may also lead to a variety of other linguistic phenomena, including language convergence, borrowing, and relexification (replacement of much of the native vocabulary with that of another language). In situations of extreme and sustained language contact, it may lead to the formation of new mixed languages that cannot be considered to belong to a single language family. One type of mixed language called pidgins occurs when adult speakers of two different languages interact on a regular basis, but in a situation where neither group learns to learn to speak the language of the other group fluently. In such a case, they will often construct a communication form that has traits of both languages, but which has a simplified grammatical and phonological structure. The language comes to contain mostly the grammatical and phonological categories that exist in both languages. Pidgin languages are defined by not having any native speakers, but only being spoken by people who have another language as their first language. But if a Pidgin language becomes the main language of a speech community, then eventually children will grow up learning the pidgin as their first language. As the generation of child learners grow up, the pidgin will often be seen to change its structure and acquire a greater degree of complexity. This type of language is generally called a creole language. An example of such mixed languages is Tok Pisin, the official language of Papua New-Guinea, which originally arose as a Pidgin based on English and Austronesian languages; others are Kreyòl ayisyen, the French based creole language spoken in Haiti, and Michif, a mixed language of Canada, based on the Native American language Cree and French.

Linguistic Diversity

A "living language" is simply one which is in wide use as a primary form of communication by a specific group of living people. The exact number of known living languages varies from 6,000 to 7,000, depending on the precision of one's definition of "language", and in particular, on how one defines the distinction between languages and dialects. As of 2009, *SIL Ethnologue* catalogued 6909 living human languages. The *Ethnologue* establishes linguistic groups based on studies of mutual intelligibility, and therefore often include more categories than more conservative classifications. For example, the Danish language that most scholars consider a single language with several dialects is classified as two distinct languages (Danish and Jutish) by the *Ethnologue.*

The *Ethnologue* is also sometimes criticized for using cumulative data gathered over many decades, meaning that exact speaker numbers are frequently out of date, and some languages classified as living may have already become extinct. According to the *Ethnologue,* 389 (or nearly 6%) languages have more than a million speakers. These languages together account for 94% of the world's population, whereas 94% of the world's languages account for the remaining 6% of the global population. To the right is a table of the world's 10 most spoken languages with population estimates from the *Ethnologue* (2009 figures).

Language Families of the World

The world's languages can be grouped into language families consisting of languages that can be shown to have common ancestry. Linguists recognise many hundreds of language families, although some of them can possibly be grouped into larger units as more evidence becomes available and in-depth studies are carried out. At present, there are also dozens of language isolates: languages that cannot be shown to be related to any other languages in the world. Among them are Basque, spoken in Europe, Zuni of New Mexico, P'urhépecha of Mexico, Ainu of Japan, Burushaski of Pakistan, and many others.

The language family of the world that has the most speakers is the Indo-European languages, spoken by 46% of the world's population. This family includes major world languages like English, Spanish, Russian, and Hindustani (Hindi/Urdu). The Indo-European family achieved prevalence first during the Eurasian Migration Period (c. 400–800 AD), and subsequently through the European colonial expansion, which brought the Indo-European languages to a politically and often numerically dominant position in the Americas and much of Africa. The Sino-Tibetan languages are spoken by 20% of the world's population and include many of the languages of East Asia, including Mandarin Chinese, Cantonese, and hundreds of smaller languages.

Africa is home to a large number of language families, the largest of which is the Niger-Congo language family, which includes such languages as Swahili, Shona, and Yoruba. Speakers of the Niger-Congo languages account for 6.9% of the world's population. A similar number of people speak the Afroasiatic languages, which include the populous Semitic languages such as Arabic, Hebrew language, and the languages of the Sahara region, such as the Berber languages and Hausa.

The Austronesian languages are spoken by 5.5% of the world's population and stretch from Madagascar to maritime Southeast Asia all the way to Oceania. It includes such languages as Malagasy, Mâori, Samoan, and many of the indigenous languages of Indonesia and Taiwan. The Austronesian languages are considered to have originated in Taiwan around 3000 BC and spread through the Oceanic region through island-hopping, based on an advanced nautical technology. Other populous language families are the Dravidian languages of South Asia (among them Tamil and Telugu), the Turkic languages of Central Asia (such as Turkish), the Austroasiatic (among them Khmer), and Tai–Kadai languages of Southeast Asia (including Thai).

The areas of the world in which there is the greatest linguistic diversity, such as the Americas, Papua New Guinea, West Africa, and South-Asia, contain hundreds of small language families. These areas together account for the majority of the world's languages, though not the majority of speakers. In the Americas, some of the largest language families include the Quechumaran, Arawak, and Tupi-Guarani families of South America, the Uto-Aztecan, Oto-Manguean, and Mayan of Mesoamerica, and the Na-Dene and Algonquian language families of North America. In Australia, most indigenous languages belong to the Pama-Nyungan family, whereas Papua-New Guinea is home to a large number of small families and isolates, as well as a number of Austronesian languages.

Speech Community

A speech community is a group of people who share a set of norms and expectations regarding the use of language.

Exactly how to define *speech community* is debated in the literature. Definitions of speech community tend to involve varying degrees of emphasis on the following:

- Shared community membership
- Shared linguistic communication

Early definitions have tended to see speech communities as bounded and localized groups of people who live together and come to share the same linguistic norms because they belong to the same local community. It has also been assumed that within a community a homogeneous set of norms should exist. These assumptions have been challenged by later scholarship that have demonstrated that individuals generally participate in various speech communities simultaneously and at different times in their lives each of which has a different norms that they tend to share only partially, communities may be de-localized and unbounded rather than local, and they often comprise different sub-communities with differing speech norms. With the recognition of the fact that speakers actively use language to construct and manipulate social identities by signalling membership in particular speech communities, the idea of the bounded speech community with homogeneous speech norms has become largely abandoned for a model based on the speech community as a fluid community of practice.

A speech community comes to share a specific set of norms for language use through living and interacting together, and speech communities may therefore emerge among all groups that interact frequently and share certain norms and ideologies. Such groups can be villages, countries, political or professional communities, communities with shared interests, hobbies, or lifestyles, or even just groups of friends. Speech communities may share both particular sets of vocabulary and grammatical conventions, as well as speech styles and genres, and also norms for how and when to speak in particular ways.

History of Definitions

The adoption of the concept of the "speech community" as a unit of linguistic analysis emerged in the 1960s.

John Gumperz

John Gumperz described how dialectologists had taken issue with the dominant approach in historical linguistics that saw linguistic communities as homogeneous and localized entities in a way that allowed for drawing neat tree diagrams based on the principle of 'descent with modification' and shared innovations. Dialectologists rather realised that dialect traits spread through diffusion and that social factors were decisive in how this happened. They also realised that traits spread as waves from centres and that often several competing varieties would exist in some communities. This insight prompted Gumperz to problematize the notion of the linguistic community as the community that carries a single speech variant, and instead to seek a definition that could encompass heterogeneity. This could be done by focusing on the interactive aspect of language, because interaction in speech is the path along which diffused linguistic traits travel. Gumperz defined the community of speech:

> Any human aggregate characterized by regular and frequent interaction by means of a shared body of verbal signs and set off from similar aggregates by significant differences in language usage. —*Gumperz (1964)*

This definition gives equal importance to the structural and interactional layers, and does not aim to delineate either the community or the language system as discrete entities. The community is a group of people that frequently interact with each other. This is not a definition of a discrete group because frequency of interaction is relative and graduated, and never stable. The definition of the language system is also not exclusive because it is defined as being set off from other systems by significant differences in usage. Furthermore Gumperz refines the definition of the linguistic system shared by a speech community:

> *Regardless of the linguistic differences among them, the speech varieties employed within a speech community form a system because they are related to a shared set of social norms.* —*Gumperz (1964)*

Here Gumperz again identifies two important components of the speech community: its members share both a set of linguistics forms and a set of social norms that govern the use of those forms. Gumperz also sought to set up a typological framework for describing how linguistic systems can be in use within a single speech community. He introduced the concept of linguistic range, the degree to which the linguistic systems of the community differ so that speech communities can be multilingual, diglossic, multidialectal (including sociolectal stratification), or homogeneous - depending on the degree of difference among the different language systems used in the community.

Secondly the notion of compartmentalization described the degree to which the use of different varieties were either set off from each other as discrete systems in interaction (e.g. diglossia where varieties correspond to specific social contexts, or multilingualism where varieties correspond to discrete social groups within the community) or whether they are habitually mixed in interaction (e.g. code-switching, bilingualism, syncretic language).

Noam Chomsky

Gumperz's formulation was however effectively overshadowed by Noam Chomsky's redefinition of the scope of linguistics as being :

> *concerned primarily with an ideal speaker-listener, in a completely homogeneous speech-community, who knows its language perfectly and is unaffected by such grammatically irrelevant conditions as memory limitations, distractions, shifts of attention and interest, and errors (random or characteristic) in applying his knowledge of the language in actual performance.*
>
> *—Chomsky (1965:3)*

Where Gumperz formulation was designed to incorporate heterogeneity, by focusing on shared norms of language use rather than a shared linguistic system, Chomsky's definition explicitly rejected it. Chomsky argued that linguistic competence was logically prior to linguistic performance, and that competence was necessarily homogeneously distributed among all speakers of a linguistic community, or language acquisition wouldn't have been possible.

William Labov

Another influential conceptualization of the linguistic community was that of William Labov, which can be seen as a hybrid of the Chomskyan structural homogeneity and Gumperz' focus on shared norms informing variable practices. Labov wrote:

> *The speech community is not defined by any marked agreement in the use of language elements, so much as by participation in a set of shared norms: these norms may be observed in overt types of evaluative behaviour, and by the uniformity of abstract patterns of variation which are invariant in respect to particular levels of usage.* *—Labov (1972:120–1)*

Like that of Gumperz, Labov's formulation stressed that a speech community was defined more by shared norms than by shared linguistic forms. But like Chomsky, Labov also saw each of the formally distinguished linguistic varieties within a speech community as homogeneous, invariant and uniform. Labov's model was designed to see speech varieties as associated with social strata within a single speech community, and it assumed each stratum to use a single variety with a well-defined, uniform structure. This model worked well for Labov's purpose which was to show that African American Vernacular

English could not be seen as structurally degenerate form of English, but rather as a well defined linguistic code with its own particular structure. Labov's model was designed to explain variation between social groups within a single speech community, and for this reason it assumed a structural integrity of the linguistic system of each social group, and it also assumed each social group within the speech community to form a neatly bounded unit definable in terms of discrete and correlatable variables, such as ethnicity, race, class, gender, age, ideology, and specific formal variables of linguistic usage.

Critique

Probably because of their considerable explanatory power, Labov's and Chomsky's understandings of the speech community became widely influential in linguistics. But gradually a number of problems with those models became apparent.

Firstly, it became increasingly clear that the assumption of homogeneity inherent in Chomsky and Labov's models was untenable. The African American speech community which Labov had seen as defined by the shared norms of AAVE, was shown to be an illusion, as ideological disagreements about the status of AAVE among different groups of speakers attracted public attention.

Secondly, in the eagerness to describe all kinds of variation in communities with a shared linguistic standard, the concept of the speech community was extended to include very large scale communities such as entire nation states, or the entire international community of English speakers. By over-extending the concept in this way Gumperz' basic requirement that the community be united by routine interaction between its members could no longer be meaningfully evoked.

Thirdly, while Chomsky and Labov's models eschewed the possibility of significant variation taking place at the level of the individual, research in interactional sociolinguistics made it increasingly clear that intra-personal variation is common. It also became clear that choice of linguistic variant is often a situational choice made in relation to a specific speech context, than it is an expression of a permanent social identity, such as class, gender, or age.

Finally, the models of speech communities that assumed a set of shared norms that differed slightly among different social classes, were criticized for assuming that each individual have equal access to all linguistic forms, but just choose to produce the kind of speech associated with their particular social group. This assumption did not take account of power differentials within the community that sometimes work to restrict individual speakers' access to speech forms of other social groups, or which impose certain linguistic varieties on certain groups and individuals.

The force of these critiques led to a general unease with the concept of *"speech communities"* because of the many contradictory connotations of the term, and because of the general turn in anthropology towards looking at social organisation in terms of hierarchy and power relations rather than studying social coherence and the construction of shared norms. Some scholars recommended abandoning the concept altogether as a preexisting object that can be studied instead conceptualizing it as "the product of the communicative activities engaged in by a given group of people." Others have proposed simply acknowledging the community's *ad hoc* status as "some kind of social group whose speech characteristics are of interest and can be described in a coherent manner".

Practice Theory

Practice theory, as developed by social thinkers such as Pierre Bourdieu, Anthony Giddens and Michel de Certeau, and especially the notion of the community of practice as developed by Jean Lave and Etienne Wenger has been influentially applied to the study of the language community by linguists such as William Hanks and Penelope Eckert.

Eckert's primary interest was in finding an approach to sociolinguistic variation that didn't presuppose any social variable as a given (e.g. class, gender, locality). Instead she aimed to build a model which was able to discover which variables are in fact the ones that matter to the group of individuals in question, the common purposes around which communities organise themselves. For Eckert the crucial defining characteristics of the community is a persistence of over time and commitment to shared understanding.

Eckert wished to focus on the subgroups and how tension between the goals and practices of subgroups that coexisting within a macro-community dynamically interrelate and generate social change. She acknowledges that Gumperz' definition of the speech community is not incompatible with the practice approach, but rather complimentary to it, and she suggests to study the two simultaneously as they mutually affect each other. Eckert's perspective on the community of practice privileges the study of how social identity is produced, and as such it studies language primarily as it relates to questions of identity.

Hanks' concept of the linguistic community as defined by linguistic practices is different from that of Eckert and Gumperz, in that rather than studying the dynamics of identity production, it studies the ways in which shared practices relate to the production of linguistic meaning. Where Eckert primarily studies how communities of practice employ linguistic practices informed by shared ideologies to demarcate themselves from other such communities, Hanks studies how linguistic practices are related to a variety of inhabitable positions within the different social fields that are constructed through shared practices.

Language Variation

The notion of speech community is most generally used as a tool to define a unit of analysis within which to analyse language variation and change. Stylistic features differ among speech communities based on factors such as the group's socioeconomic status, common interests and the level of formality expected within the group and by its larger society.

In Western culture, for example, employees at a law office would likely use more formal language than a group of teenage skateboarders because most Westerners expect more formality and professionalism from practitioners of law than from an informal circle of adolescent friends. This special use of language by certain professions for particular activities is known in linguistics as register; in some analyses, the group of speakers of a register is known as a discourse community, while the phrase "speech community" is reserved for varieties of a language or dialect that speakers inherit by birth or adoption.

Chapter 2

Caste and Religion in Society

Caste

Caste is a form of social stratification characterized by endogamy, hereditary transmission of a lifestyle which often includes an occupation, ritual status in a hierarchy and customary social interaction and exclusion based on cultural notions of purity and pollution. According to Human Rights Watch and UNICEF, caste discrimination affects an estimated 250 million people worldwide.

A paradigmatic, ethnographic example is the division of Indian society into rigid social groups, with roots in India's ancient history and persisting until today. Historically, the caste system in India has consisted of thousands of endogamous groups called Jatis or Quoms (among Muslims). Independent India has witnessed caste-related violence. The Nepalese caste system resembles that of the Indian Jâti system with numerous Jâti divisions with a Varna system superimposed for a rough equivalence. Religious, historical and sociocultural factors have helped define the bounds of endogamy for Muslims in some parts of Pakistan. The Caste system in Sri Lanka is a division of society into strata, influenced by the classic Aryan Varnas of North India and the Dravida Jâti system found in South India.

Balinese caste structure has been described in early 20th-century European literature to be based on three categories – triwangsa (thrice born) or the nobility, dwijati (twice born) in contrast to ekajati (once born) the low folks. In China during the period of Yuan Dynasty, ruler Kublai Khan enforced a *Four Class System*, which was a legal caste system. The order of four classes of people was maintained by the information of the descending order were: Mongolian, Semu people,

Han people (in the northern areas of China), and Southerners (people of the former Southern Song Dynasty). In Japan's history, social strata based on inherited position rather than personal merits, was rigid and highly formalized. With the unification of the three kingdoms in the 7th century and the foundation of the Goryeo dynasty in the Middle Ages, Koreans systemised its own native class system.

Yezidi society is hierarchical. In Yemen there exists a hereditary caste, the African-descended Al-Akhdam who are kept as perennial manual workers. Various sociologists have reported caste systems in Africa.

Caste in South Asia

Caste System of India: Historically, the caste system in India has consisted of thousands of endogamous groups called Jatis or Quoms (among Muslims). All the Jatis were clubbed under the *varnas* categories during the British colonial Census of 1901. The terms *varna* (theoretical classification based on occupation) and *jâti* (caste) are two distinct concepts: while varna is the idealised four-part division envisaged by the Twice-Borns, jâti (community) refers to the thousands of actual endogamous groups prevalent across the subcontinent. A jati may be divided into exogamous groups based on same gotras. The classical authors scarcely speak of anything other than the varnas; even Indologists sometimes confuse the two.

Independent India has witnessed caste-related violence. In 2005, government statistics recorded approximately 110,000 cases of reported violent acts, including rape and murder, committed against Dalits The economic significance of the caste system in India has been declining as a result of urbanization and affirmative action programs. Upon independence from the British rule, the Indian Constitution listed 1,108 castes across the country as Scheduled Castes in 1950, for affirmative action. The Scheduled Castes are sometimes called Dalit in contemporary literature. In 2001, the proportion of Dalit population was 16.2 percent of India's total population. The majority of the 15 million bonded child workers in India, are from the lowest castes.

Nepal: The Nepalese caste system resembles that of the Indian Jâti system with numerous Jâti divisions with a Varna system superimposed for a rough equivalence. But since the culture and the society is different some of the things are different. Inscriptions attest the beginnings of a caste system during the Lichchhavi period. Jayasthiti Malla (1382–95) categorized Newars into 64 castes (Gellner 2001). A similar exercise was made during the reign of Mahindra Malla (1506–75). The Hindu social code was later set up in Gorkha by Ram Shah (1603–36).

Pakistan: Religious, historical and sociocultural factors have helped define the bounds of endogamy for Muslims in some parts of Pakistan. There is a preference for endogamous marriages based on the clan-oriented nature of the society, which values and actively seeks similarities in social group identity based on several factors, including religious, sectarian, ethnic, and tribal/clan affiliation. Religious affiliation is itself multilayered and includes religious considerations other than being Muslim, such as sectarian identity (e.g. Shia or Sunni, etc.) and religious orientation within the sect (Isnashari, Ismaili, Ahmedi, etc.).

Both ethnic affiliation (e.g. Pathan, Sindhi, Baloch, Punjabi, etc.) and membership of specific biraderis or zaat/quoms are additional integral components of social identity. Within the bounds of endogamy defined by the above parameters, close consanguineous unions are preferred due to a congruence of key features of group- and individual-level background factors as well as affinities. McKim Marriott claims a social stratification that is hierarchical, closed, endogamous and hereditary is widely prevalent, particularly in western parts of Pakistan. Frederik Barth in his review of this system of social stratification in Pakistan suggested that these are castes.

Sri Lanka: The Caste system in Sri Lanka is a division of society into strata, influenced by the classic Aryan Varnas of North India and the Dravida Jâti system found in South India. Ancient Sri Lankan texts such as the Pujavaliya, Sadharmaratnavaliya and Yogaratnakaraya and inscriptional evidence show that the above hierarchy prevailed throughout the feudal period. The repetition of the same caste hierarchy even as recently as the 18th century, in the British/Kandyan period Kadayimpoth - Boundary books as well, indicates the continuation of the tradition right up to the end of Sri Lanka's monarchy.

Caste-like Stratification Outside South Asia

Southeast Asia:

Indonesia: Balinese caste structure has been described in early 20th-century European literature to be based on three categories – triwangsa (thrice born) or the nobility, dwijati (twice born) in contrast to ekajati (once born) the low folks. Four statuses were identified in these sociological studies, spelled a bit differently from the caste categories for India:

- Brahmanas - priest
- Satrias - knighthood

- Wesias - commerce
- Sudras - servitude

The Brahmana caste was further subdivided by these Dutch ethnographers into two: Siwa and Buda. The Siwa caste was subdivided into five – Kemenuh, Keniten, Mas, Manuba and Petapan. This classification was to accommodate the observed marriage between higher caste Brahmana men with lower caste women. The other castes were similarly further sub-classified by these 19th-century and early-20th-century ethnographers based on numerous criteria ranging from profession, endogamy or exogamy or polygamy, and a host of other factors in a manner similar to *castas* in Spanish colonies such as Mexico, and caste system studies in British colonies such as India.

East Asia:

China and Mongolia: During the period of Yuan Dynasty, ruler Kublai Khan enforced a *Four Class System*, which was a legal caste system. The order of four classes of people was maintained by the information of the descending order were:-

- Mongolian
- Semu people
- Han people (in the northern areas of China)
- Southerners (people of the former Southern Song Dynasty)

Some scholars notes that it was a kind of psychological indication that the earlier they submitted to Mongolian people, the higher social status they would have. The 'Four Class System' and its people received different treatment in political, legal, and military affairs.

Today, the Hukou system is considered by various sources as the current caste system of the China.

Japan: In Japan's history, social strata based on inherited position rather than personal merits, was rigid and highly formalized. At the top were the Emperor and Court nobles (kuge), together with the Shogun and daimyo. Below them the population was divided into four classes in a system known as *mibunsei* («ŽR6R). These were: samurai, peasants, craftsmen and merchants. Only the samurai class was allowed to bear arms. A samurai had a right to kill any peasants and other craftsmen and merchants whom he felt were disrespectful. Craftsmen produced products, being the third, and the last merchants were thought to be as the meanest class because they did not produce any products. The castes were further sub-divided; for example, the peasant caste were labelled as *furiuri*, *tanagari*, *mizunomi-byakusho* amongst others. The castes and sub-classes, as in Europe, were from the same race, religion and culture.

Howell, in his review of Japanese society notes that if a Western power had colonized Japan in the 19th century, they would have discovered and imposed a rigid four-caste hierarchy in Japan.

De Vos and Wagatsuma observe that a systematic and extensive caste system was part of the Japanese society. They also discuss how alleged caste impurity and alleged racial inferiority, concepts often quickly assumed to be slightly different, are superficial terms, two faces of identical inner psychological processes, which expressed themselves in Japan and other countries of the world.

Endogamy was common because marriage across caste lines was socially unacceptable.

Japan had its own untouchable caste, shunned and ostracized, historically referred to by the insulting term *Eta*, now called *Burakumin*. While modern law has officially abolished the class hierarchy, there are reports of discrimination against the Buraku or Burakumin underclasses. The Burakumin are regarded as "ostracised." The burakumin are one of the main minority groups in Japan, along with the Ainu of Hokkaidô and those of residents of Korean and Chinese descent.

Korea: With the unification of the three kingdoms in the 7th century and the foundation of the Goryeo dynasty in the Middle Ages, Koreans systemised its own native class system. At the top were the two official classes, the Yangban that literally means "two classes." It was composed of scholars (Munban) and warriors (Muban). Within the Yangban class, the Scholars (Munban) enjoyed a significant social advantage over the warrior (Muban) class, until the Muban Rebellion in 1170 resulting in the 100 year Goryeo military regime. Muban ruled Korea under successive Warrior Leaders until the final Mongol victory in 1270. In 1392, with the foundation of Confucian Joseon dynasty, the full ascendancy of munban over muban was final.

Beneath the Yangban class were the *Jung-in* (ÉxÇ--N°N: literally "middle people"). They were the technicians. This class was small and specialised in fields such as medicine, accounting, translators, regional bureaucrats, etc.

Beneath the Jung-in were the *Sangmin* (ÁÀü»-8^l: literally 'commoner'). These were independent farmers working their own fields.

Underneath them all were the Baekjeong. The meaning today is that of butcher. They originate from the Khitan invasion of Korea in the 11th century. The defeated Khitans who had surrendered were settled in isolated communities throughout Goryeo to forestall rebellion. They were valued for their skills in hunting, herding, butchering, and making of leather, common skill sets among nomads. Over time their ethnic origin was forgotten, and they formed the bottom layer of Korean society.

Korea had a very large slave population, *nobi*, ranging from a third to half of the entire population for most of the millennium between the Silla period and the Joseon Dynasty. Slavery was legally abolished in Korea in 1894 but remained extant in reality until 1930.

The opening of Korea to foreign Christian missionary activity in the late 19th century saw some improvement in the status of the *baekjeong*; However, everyone was not equal under the Christian congregation, and protests erupted when missionaries attempted to integrate them into worship services, with non-*baekjeong* finding such an attempt insensitive to traditional notions of hierarchical advantage. Also around the same time, the baekjeong began to resist the open social discrimination that existed against them. They focused on social and economic injustices affecting the baekjeong, hoping to create an egalitarian Korean society. Their efforts included attacking social discrimination by the upper class, authorities, and "commoners" and the use of degrading language against children in public schools.

With the Gabo reform of 1896, the class system of Korea was officially abolished. However, the Yangban families carried on traditional education and formal mannerisms into the 21st century. Gabo reform of 1896 required all peasants to adopt a last name. The vast majority of peasants who did not possess a last name opted to adopt the most prestigious of family names, resulting in the lopsided presence of Kim, Lee, and Park names among Koreans. These newly minted families do not possess Clan affiliations or genealogies. The Yangban families of Korea maintain their lineage through centuries old genealogy passed down from generation to generation. Most Yangban Koreans can name their family clan, one level beyond their last name, and possess a character in his first name containing a hereditary generation name, called dollimja (Ì³¼¹□Ç) or hangryeolja (mÕ,¸□Ç) in Korean.

North Korea: Committee for Human Rights in North Korea reported that "Every North Korean citizen is assigned a heredity-based class and socio-political rank over which the individual exercises no control but which determines all aspects of his or her life." Regarded as Songbun, Barbara Demick describes this "class structure" as an updating of the hereditary "caste system", combining Confucianism and Stalinism. She claims that a bad family background is called "tainted blood", and that by law this "tainted blood" lasts for three generations.

West Asia

Yezidi society is hierarchical. The secular leader is a hereditary emir or prince, whereas a chief sheikh heads the religious hierarchy. The Yazidi are strictly endogamous; members of the three Yazidi castes, the murids, sheikhs and pirs, marry only within their group.

Yemen: In Yemen there exists a hereditary caste, the African-descended Al-Akhdam who are kept as perennial manual workers. Estimates put their number at over 3.5 million residents who are discriminated, out of a total Yemeni population of around 22 million.

Africa

Various sociologists have reported caste systems in Africa. The specifics of the caste systems have varied in ethnically and culturally diverse Africa, however the following features are common - it has been a closed system of social stratification, the social status is inherited, the castes are hierarchical, certain castes are shunned while others are merely endogamous and exclusionary. In some cases, concepts of purity and impurity by birth have been prevalent in Africa. In other cases, such as the *Nupe* of Nigeria, the *Beni Amer* of East Africa, and the *Tira* of Sudan, the exclusionary principle has been driven by evolving social factors.

West Africa

Among the Igbo of Nigeria - especially Enugu, Anambra, Imo, Abia, Ebonyi, Edo and Delta states of the country - Obinna finds Osu caste system has been and continues to be a major social issue. The Osu caste is determined by one's birth into a particular family irrespective of the religion practised by the individual. Once born into Osu caste, this Nigerian person is an outcast, shunned and ostracized, with limited opportunities or acceptance, regardless of his or her ability or merit. Obinna discusses how this caste system-related identity and power is deployed within government, Church and indigenous communities.

The *osu* class systems of eastern Nigeria and southern Cameroon are derived from indigenous religious beliefs and discriminate against the "Osus" people as "owned by deities" and outcasts.

The Songhai economy was based on a caste system. The most common were metalworkers, fishermen, and carpenters. Lower caste participants consisted of mostly non-farm working immigrants, who at times were provided special privileges and held high positions in society. At the top were noblemen and direct descendants of the original Songhai people, followed by freemen and traders.

In a review of social stratification systems in Africa, Richter reports that the term caste has been used by French and American scholars to many groups of West African artisans. These groups have been described as inferior, deprived of all political power, have a specific occupation, are hereditary and sometimes despised by others. Richter illustrates caste system in Cote d'lvoire, with six sub-caste categories. Unlike other parts of the world, mobility is sometimes possible within sub-castes, but not across caste lines. Farmers and artisans have been, claims Richter, distinct castes. Certain sub-castes are shunned more than others. For example, exogamy is rare for women born into families of woodcarvers.

Similarly, the Mandé societies in Gambia, Ghana, Guinea, Ivory Coast, Liberia, Senegal and Sierra Leone have social stratification systems that divide society by ethnic ties. The Mande class system regards the *jonow* slaves as inferior. Similarly, the Wolof in Senegal is divided into three main groups, the *geer* (freeborn/nobles), *jaam* (slaves and slave descendants) and the underclass *neeno*. In various parts of West Africa, Fulani societies also have class divisions. Other castes include *Griots*, *Forgerons*, and *Cordonniers*.

Tamari has described endogamous castes of over fifteen West African peoples, including the Tukulor, Songhay, Dogon, Senufo, Minianka, Moors, Manding, Soninke, Wolof, Serer, Fulani, and Tuareg. Castes appeared among the *Malinke* people no later than 14th century, and was present among the *Wolof* and *Soninke*, as well as some *Songhay* and *Fulani* populations, no later than 16th century. Tamari claims that wars, such as the *Sosso-Malinke* war described in the *Sunjata* epic, led to the formation of blacksmith and bard castes among the people that ultimately became the Mali empire.

As West Africa evolved over time, sub-castes emerged that acquired secondary specialisations or changed occupations. Endogamy was prevalent within a caste or among a limited number of castes, yet castes did not form demographic isolates according to Tamari. Social status according to caste was inherited by off-springs automatically; but this inheritance was paternal. That is, children of higher caste men and lower caste or slave concubines would have the caste status of the father.

Central Africa

Ethel M. Albert in 1960 claimed that the societies in Central Africa were caste-like social stratification systems. Similarly, in 1961, Maquet notes that the society in Rwanda and Burundi can be best described as castes. The Tutsi, noted Maquet, considered themselves as superior, with the more numerous Hutu and the least numerous Twa regarded, by birth, as respectively, second and third in the hierarchy of Rwandese society. These groups were largely endogamous, exclusionary and with limited mobility. Maquet's theories have been controversial.

Horn of Africa

In a review published in 1977, Todd reports that numerous scholars report a system of social stratification in different parts of Africa that resembles some or all aspects of caste system. Examples of such caste systems, he claims, are to be found in Ethiopia in communities such as the Gurage and Konso. He then presents the Dime of Southwestern Ethiopia, amongst whom there operates a system which Todd claims can be unequivocally labelled as caste system. The Dime have seven castes whose size varies considerably. Each broad caste level is a hierarchical order that is based on notions of purity, non-purity and impurity. It uses the concepts of defilement to limit contacts between caste categories and to preserve the purity of the upper castes. These caste categories have been exclusionary, endogamous and the social identity inherited. Alula Pankhurst has published a study of caste groups in SW Ethiopia.

Among the Kafa, there were also traditionally groups labelled as castes. "Based on research done before the Derg regime, these studies generally presume the existence of a social hierarchy similar to the caste system. At the top of this hierarchy were the Kafa, followed by occupational groups including blacksmiths (Qemmo), weavers (Shammano), bards (Shatto), potters, and tanners (Manno). In this hierarchy, the Manjo were commonly referred to as hunters, given the lowest status equal only to slaves."

The Borana Oromo of southern Ethiopia in the Horn of Africa also have a class system, wherein the Wata, an acculturated hunter-gatherer group, represent the lowest class. Though the Wata today speak the Oromo language, they have traditions of having previously spoken another language before adopting Oromo.

The traditionally nomadic Somali people are divided into clans, wherein the Rahanweyn agro-pastoral clans and the occupational clans such as the Madhiban were traditionally sometimes treated as outcasts. As Gabboye, the Madhiban along with the Yibir and Tumaal (collectively referred to as *sab*) have since obtained political representation within Somalia, and their general social status has improved with the expansion of urban centres.

Europe

France and Spain: For centuries, through the modern times, the majority regarded Cagots of western France and northern Spain as an inferior caste, the untouchables. While they had the same skin colour and religion as the majority, in the Churches, they had to use segregated doors, drink from segregated fonts, receive communion on the end of long wooden spoons. It was a closed social system. The socially isolated Cagots were endogamous, and chances of social mobility non-existent.

Religion

Figure: *Religious activities around the world*

A religion is an organised collection of beliefs, cultural systems, and world views that relate humanity to an order of existence. Many religions have narratives, symbols, and sacred histories that are intended to explain the meaning of life and/or to explain the origin of life or the Universe. From their beliefs about the cosmos and human nature, people derive morality, ethics, religious laws or a preferred lifestyle. According to some estimates, there are roughly 4,200 religions in the world.

Many religions may have organised behaviours, clergy, a definition of what constitutes adherence or membership, holy places, and scriptures. The practice of a religion may also include rituals, sermons, commemoration or veneration of a deity, gods or goddesses, sacrifices, festivals, feasts, trance, initiations, funerary services, matrimonial services, meditation, prayer, music, art, dance, public service or other aspects of human culture. Religions may also contain mythology. The word *religion* is sometimes used interchangeably with *faith, belief system* or sometimes *set of duties*; however, in the words of Émile Durkheim, religion differs from private belief in that it is "something eminently social". A global 2012 poll reports that 59% of the world's population is religious, and 36% are not religious, including 13% who are atheists, with a 9% decrease in religious belief from 2005. On average, women are more religious than men. Some people follow multiple religions or multiple religious principles at the same time, regardless of whether or not the religious principles they follow traditionally allow for syncretism.

Definitions

There are numerous definitions of religion and only a few are stated here. The typical dictionary definition of religion refers to a "belief in, or the worship of, a god or gods" or the "service and worship of God or the supernatural". However, writers and scholars have expanded upon the "belief in god" definitions as insufficient to capture the diversity of religious thought and experience.

Peter Mandaville and Paul James define religion as "a relatively-bounded system of beliefs, symbols and practices that addresses the nature of existence, and in which communion with others and Otherness is *lived* as if it both takes in and spiritually transcends socially-grounded ontologies of time, space, embodiment and knowing". This definition is intended, they write, to get away from the modernist dualisms or dichotomous understandings of immanence/transcendence, spirituality/materialism, and sacredness/secularity.

Edward Burnett Tylor defined religion as "the belief in spiritual beings". He argued, back in 1871, that narrowing the definition to mean the belief in a supreme deity or judgment after death or idolatry and so on, would exclude many peoples from the category of religious, and thus "has the fault of identifying religion rather with particular developments than with the deeper motive which underlies them". He also argued that the belief in spiritual beings exists in all known societies.

The anthropologist Clifford Geertz defined religion as a "system of symbols which acts to establish powerful, pervasive, and long-lasting moods and motivations in men by formulating conceptions of a general order of existence and clothing these conceptions with such an aura of factuality that the moods and motivations seem uniquely realistic." Alluding perhaps to Tylor's "deeper motive", Geertz remarked that "we have very little idea of how, in empirical terms, this particular miracle is accomplished. We just know that it is done, annually, weekly, daily, for some people almost hourly; and we have an enormous ethnographic literature to demonstrate it". The theologian Antoine Vergote also emphasized the "cultural reality" of religion, which he defined as "the entirety of the linguistic expressions, emotions and, actions and signs that refer to a supernatural being or supernatural beings"; he took the term "supernatural" simply to mean whatever transcends the powers of nature or human agency.

The sociologist Durkheim, in his seminal book The Elementary Forms of the Religious Life, defined religion as a "unified system of beliefs and practices relative to sacred things". By sacred things he meant things "set apart and forbidden—beliefs and practices which unite into one single moral community called a Church, all those who adhere to them". Sacred things are not, however, limited to gods or spirits. On the contrary, a sacred thing can be "a rock, a tree, a spring, a pebble, a piece of wood, a house, in a word, anything can be sacred". Religious beliefs, myths, dogmas and legends are the representations that express the nature of these sacred things, and the virtues and powers which are attributed to them.

In his book The Varieties of Religious Experience, the psychologist William James defined religion as "the feelings, acts, and experiences of individual men in their solitude, so far as they apprehend themselves to stand in relation to whatever they may consider the divine". By the term "divine" James meant "any object that is god*like*, whether it be a concrete deity or not" to which the individual feels impelled to respond with solemnity and gravity.

Echoes of James' and Durkheim's definitions are to be found in the writings of, for example, Frederick Ferré who defined religion as "one's way of valuing most comprehensively and intensively". Similarly, for the theologian Paul Tillich, faith is "the state of being ultimately concerned", which "is itself religion. Religion is the substance, the ground, and the depth of man's spiritual life." Friedrich Schleiermacher in the late 18th century defined religion as *das schlechthinnige Abhängigkeitsgefühl*, commonly translated as "a feeling of absolute dependence". His contemporary Hegel disagreed thoroughly, defining religion as "the Divine Spirit becoming conscious of Himself through the finite spirit."

When religion is seen in terms of "sacred", "divine", intensive "valuing", or "ultimate concern", then it is possible to understand why scientific findings and philosophical criticisms (e.g. Richard Dawkins) do not necessarily disturb its adherents. An increasing number of scholars have expressed reservations about ever defining the 'essence' of religion. They observe that the way we use the concept today is a particularly modern construct that would not have been understood through much of history and in many cultures outside the west (or even in the west until after the Peace of Westphalia).

Theories

Origins and Development: The origin of religion is uncertain. There are a number of theories regarding the subsequent origins of organised religious practices.

According to anthropologists John Monaghan and Peter Just, "Many of the great world religions appear to have begun as revitalization movements of some sort, as the vision of a charismatic prophet fires the imaginations of people seeking a more comprehensive answer to their problems than they feel is provided by everyday beliefs. Charismatic individuals have emerged at many times and places in the world. It seems that the key to long-term success – and many movements come and go with little long-term effect – has relatively little to do with the prophets, who appear with surprising regularity, but more to do with the development of a group of supporters who are able to institutionalize the movement."

The development of religion has taken different forms in different cultures. Some religions place an emphasis on belief, while others emphasize practice. Some religions focus on the subjective experience of the religious individual, while others consider the activities of the religious community to be most important. Some religions claim to be universal, believing their laws and cosmology to be binding for everyone, while others are intended to be practiced only by a closely defined or localized group. In many places religion has been associated with public institutions such as education, hospitals, the family, government, and political hierarchies. Anthropologists John Monoghan and Peter Just state that, "it seems apparent that one thing religion or belief helps us do is deal with problems of human life that are significant, persistent, and intolerable. One important way in which religious beliefs accomplish this is by providing a set of ideas about how and why the world is put together that allows people to accommodate anxieties and deal with misfortune."

Social Constructionism

One modern academic theory of religion, social constructionism, says that religion is a modern concept that suggests all spiritual practice and worship follows a model similar to the Abrahamic religions as an orientation system that helps to interpret reality and define human beings. Among the main proponents of this theory of religion are Daniel Dubuisson, Timothy Fitzgerald, Talal Asad, and Jason Ânanda Josephson. The social constructionists argue that religion is a modern concept that developed from Christianity and was then applied inappropriately to non-Western cultures.

Daniel Dubuisson, a French anthropologist, says that the idea of religion has changed a lot over time and that one cannot fully understand its development by relying on consistent use of the term, which "tends to minimize or cancel out the role of history". "What the West and the history of religions in its wake have objectified under the name 'religion'", he says, " is ... something quite unique, which could be appropriate only to itself and its own history." He notes that St. Augustine's definition of *religio* differed from the way we used the modern word "religion".

Dubuisson prefers the term "cosmographic formation" to religion. Dubuisson says that, with the emergence of religion as a category separate from culture and society, there arose religious studies. The initial purpose of religious studies was to demonstrate the superiority of the "living" or "universal" European world view to the "dead" or "ethnic" religions scattered throughout the rest of the world, expanding the teleological project of Schleiermacher and Tiele to a worldwide ideal religiousness. Due to shifting theological currents, this was eventually supplanted by a liberal-ecumenical interest in searching for Western-style universal truths in every cultural tradition.

According to Fitzgerald, religion is not a universal feature of all cultures, but rather a particular idea that first developed in Europe under the influence of Christianity. Fitzgerald argues that from about the 4th century CE Western Europe and the rest of the world diverged. As Christianity became commonplace, the charismatic authority identified by Augustine, a quality we might today call "religiousness", exerted a commanding influence at the local level. As the Church lost its dominance during the Protestant Reformation and Christianity became closely tied to political structures, religion was recast as the basis of national sovereignty, and religious identity gradually became a less universal sense of spirituality and more divisive, locally defined, and tied to nationality. It was at this point that "religion" was dissociated with universal beliefs and moved closer to dogma in both meaning and practice. However there was not yet the idea of dogma as a personal choice, only of established churches. With the Enlightenment religion lost its attachment to nationality, says Fitzgerald, but rather than becoming a universal social attitude, it now became a personal feeling or emotion.

Asad argues that before the word "religion" came into common usage, Christianity was a *disciplina*, a "rule" just like that of the Roman Empire. This idea can be found in the writings of St. Augustine (354–430). Christianity was then a power structure opposing and superseding human institutions, a literal Kingdom of Heaven. It was the discipline taught by one's family, school, church, and city authorities, rather than something calling one to self-discipline through symbols.

These ideas are developed by S. N. Balagangadhara. In the Age of Enlightenment, Balagangadhara says that the idea of Christianity as the purest expression of spirituality was supplanted by the concept of "religion" as a worldwide practice. This caused such ideas as religious freedom, a re-examination of classical philosophy as an alternative to Christian thought, and more radically Deism among intellectuals such as Voltaire. Much like Christianity, the idea of "religious freedom" was exported around the world as a civilizing technique, even to regions such as India that had never treated spirituality as a matter of political identity.

More recently, in *The Invention of Religion in Japan*, Josephson has argued that while the concept of "religion" was Christian in its early formulation, non-Europeans (such as the Japanese) did not just acquiesce and passively accept the term's meaning. Instead they worked to interpret "religion" (and its boundaries) strategically to meet their own agendas and staged these new meanings for a global audience. In nineteenth century Japan, Buddhism was radically transformed from a pre-modern philosophy of natural law into a "religion," as Japanese leaders worked to address domestic and international political concerns. In summary, Josephson argues that the European encounter with other cultures has led to a partial de-Christianization of the category religion. Hence "religion" has come to refer to a confused collection of traditions with no possible coherent definition.

George Lindbeck, a Lutheran and a postliberal theologian (but not a social constructionist), says that religion does not refer to belief in "God" or a transcendent Absolute, but rather to "a kind of cultural and/or linguistic framework or medium that shapes the entirety of life and thought ... it is similar to an idiom that makes possible the description of realities, the formulation of beliefs, and the experiencing of inner attitudes, feelings, and sentiments."

Comparative Religion

Nicholas de Lange, Professor of Hebrew and Jewish Studies at Cambridge University, says that "The comparative study of religions is an academic discipline which has been developed within Christian theology faculties, and it has a tendency to force widely differing phenomena into a kind of strait-jacket cut to a Christian pattern. The problem is not only that other 'religions' may have little or nothing to say about questions which are of burning importance for Christianity, but that they may not even see themselves as religions in precisely the same way in which Christianity sees itself as a religion."

Types

Categories: Some scholars classify religions as either *universal religions* that seek worldwide acceptance and actively look for new converts, or *ethnic religions* that are identified with a particular ethnic group and do not seek converts. Others reject the distinction, pointing out that all religious practices, whatever their philosophical origin, are ethnic because they come from a particular culture.

In the 19th and 20th centuries, the academic practice of comparative religion divided religious belief into philosophically defined categories called "world religions." However, some recent scholarship has argued that not all types of religion are necessarily separated by mutually exclusive philosophies, and furthermore that the utility of ascribing a practice to a certain philosophy, or even calling a given practice religious, rather than cultural, political, or social in nature, is limited. The current state of psychological study about the nature of religiousness suggests that it is better to refer to religion as a largely invariant phenomenon that should be distinguished from cultural norms (i.e. "religions").

Some academics studying the subject have divided religions into three broad categories:

1. world religions, a term which refers to transcultural, international faiths;
2. indigenous religions, which refers to smaller, culture-specific or nation-specific religious groups; and
3. new religious movements, which refers to recently developed faiths.

Interfaith Cooperation

Because religion continues to be recognised in Western thought as a universal impulse, many religious practitioners have aimed to band together in interfaith dialogue, cooperation, and religious peacebuilding. The first major dialogue was the Parliament of the World's Religions at the 1893 Chicago World's Fair, which remains notable even today both in affirming "universal values" and recognition of the diversity of practices among different cultures. The 20th century has been especially fruitful in use of interfaith dialogue as a means of solving ethnic, political, or even religious conflict, with Christian–Jewish reconciliation representing a complete reverse in the attitudes of many Christian communities towards Jews.

Recent interfaith initiatives include "A Common Word", launched in 2007 and focused on bringing Muslim and Christian leaders together, the "C1 World Dialogue", the "Common Ground" initiative between Islam and Buddhism, and a United Nations sponsored "World Interfaith Harmony Week".

Religious Groups

The list of still-active religious movements given here is an attempt to summarize the most important regional and philosophical influences on local communities, but it is by no means a complete description of every religious community, nor does it explain the most important elements of individual religiousness.

The five largest religious groups by world population, estimated to account for 5 billion people, are Christianity, Islam, Buddhism, Hinduism (with the relative numbers for Buddhism and Hinduism dependent on the extent of syncretism) and Chinese folk religion.

Five largest religions	*2012 (billion)*	*2012 (%)*	*2000 (billion)*	*2000 (%)*	*Demographics*
Christianity	2.2	32%	2.0	33%	Christianity by country
Islam	1.6	23%	1.2	19.6%	Islam by country
Hinduism	1.0	15%	0.811	13.4%	Hinduism by country
Buddhism	0.5	7%	0.360	5.9%	Buddhism by country
Chinese folk religion	0.4	6%	0.385	6.4%	Chinese folk religion
Total	6.9	100%	6.15	100%	

Abrahamic

Abrahamic religions are monotheistic religions which believe they descend from Abraham.

- Judaism is the oldest Abrahamic religion, originating in the people of ancient Israel and Judea. Judaism is based primarily on the Torah, a text which some Jews believe was handed down to the people of Israel through the prophet Moses. This along with the rest of the Hebrew Bible and the Talmud are the central texts of Judaism. The Jewish people were scattered after the destruction of the Temple in Jerusalem in 70 CE. Today there are about 13 million Jews, about 40 per cent living in Israel and 40 per cent in the United States.

- Christianity is based on the life and teachings of Jesus of Nazareth (1st century) as presented in the New Testament. The Christian faith is essentially faith in Jesus as the Christ, the Son of God, and as Savior and Lord. Almost all Christians believe in the Trinity, which teaches the unity of Father, Son (Jesus Christ), and Holy Spirit as three persons in one Godhead. Most Christians can describe their faith with the Nicene Creed. As the religion of Byzantine Empire in the first millennium and of Western Europe during the time of colonization, Christianity has been propagated throughout the world. The main divisions of Christianity are, according to the number of adherents:
    - Catholic Church, headed by the Pope in Rome, is a communion of the Western church and 22 Eastern Catholic churches.
    - Protestantism, separated from the Catholic Church in the 16th-century Reformation and split in many denominations,
    - Eastern Christianity, which include Eastern Orthodoxy, Oriental Orthodoxy, and the Church of the East.

There are other smaller groups, such as Jehovah's Witnesses and the Latter Day Saint movement, whose inclusion in Christianity is sometimes disputed.

- Islam is based on the Quran, one of the holy books considered by Muslims to be revealed by God, and on the teachings (hadith) of the Islamic prophet Muhammad, a major political and religious figure of the 7th century CE. Islam is the most widely practiced religion of Southeast Asia, North Africa, Western Asia, and Central Asia, while Muslim-majority countries also exist in parts of South Asia, Sub-Saharan Africa, and Southeast Europe. There are also several Islamic republics, including Iran, Pakistan, Mauritania, and Afghanistan.
    - Sunni Islam is the largest denomination within Islam and follows the Quran, the hadiths which record the sunnah, whilst placing emphasis on the sahabah.
    - Shia Islam is the second largest denomination of Islam and its adherents believe that Ali succeeded Muhammad and further places emphasis on Muhammad's family.

  - o Ahmadiyya adherents believe that the awaited Imam Mahdi and the Promised Messiah has arrived, believed to be Mirza Ghulam Ahmad by Ahmadis.
  - o Other denominations of Islam include Nation of Islam, Ibadi, Sufism, Quranism, Mahdavia, and non-denominational Muslims. Wahhabism is the dominant Muslim schools of thought in the Kingdom of Saudi Arabia.
- The Bahá'í Faith is an Abrahamic religion founded in 19th century Iran and since then has spread worldwide. It teaches unity of all religious philosophies and accepts all of the prophets of Judaism, Christianity, and Islam as well as additional prophets including its founder Bahá'u'lláh.
- Smaller regional Abrahamic groups, including Samaritanism (primarily in Israel and the West Bank), the Rastafari movement (primarily in Jamaica), and Druze (primarily in Syria and Lebanon).

Iranian

Iranian religions are ancient religions whose roots predate the Islamization of Greater Iran. Nowadays these religions are practiced only by minorities.

- Zoroastrianism is a religion and philosophy based on the teachings of prophet Zoroaster in the 6th century BC. The Zoroastrians worship the Creator Ahura Mazda. In Zoroastrianism good and evil have distinct sources, with evil trying to destroy the creation of Mazda, and good trying to sustain it.
- Mandaeism is a monotheistic religion with a strongly dualistic worldview. Mandaeans are sometime labelled as the "Last Gnostics".
- Kurdish religions include the traditional beliefs of the Yazidi, Alevi, and Ahl-e Haqq. Sometimes these are labelled Yazdânism.

Indian

Indian religions are practiced or were founded in the Indian subcontinent. They are sometimes classified as the *dharmic religions*, as they all feature dharma, the specific law of reality and duties expected according to the religion.

- Hinduism is a synecdoche describing the similar philosophies of Vaishnavism, Shaivism, and related groups practiced or founded in the Indian subcontinent. Concepts most of them share in common include karma, caste, reincarnation, mantras, yantras, and darœana. Hinduism is the most ancient of still-active religions, with origins perhaps as far back as prehistoric times. Hinduism is not a monolithic religion but a religious category containing dozens of separate philosophies amalgamated as Sanâtana Dharma, which is the name with whom Hinduism has been known throughout history by its followers.

- Jainism, taught primarily by Parsva (9th century BCE) and Mahavira (6th century BCE), is an ancient Indian religion that prescribes a path of non-violence for all forms of living beings in this world. Jains are found mostly in India.
- Buddhism was founded by Siddhartha Gotama in the 6th century BCE. Buddhists generally agree that Gotama aimed to help sentient beings end their suffering (dukkha) by understanding the true nature of phenomena, thereby escaping the cycle of suffering and rebirth (saCsâra), that is, achieving nirvana.
  - Theravada Buddhism, which is practiced mainly in Sri Lanka and Southeast Asia alongside folk religion, shares some characteristics of Indian religions. It is based in a large collection of texts called the Pali Canon.
  - Mahayana Buddhism (or the "Great Vehicle") under which are a multitude of doctrines that became prominent in China and are still relevant in Vietnam, Korea, Japan and to a lesser extent in Europe and the United States. Mahayana Buddhism includes such disparate teachings as Zen, Pure Land, and Soka Gakkai.
  - Vajrayana Buddhism first appeared in India in the 3rd century CE. It is currently most prominent in the Himalaya regions and extends across all of Asia (cf. Mikkyô).
  - Two notable new Buddhist sects are Hòa H£o and the Dalit Buddhist movement, which were developed separately in the 20th century.
- Sikhism is a monotheistic religion founded on the teachings of Guru Nanak and ten successive Sikh gurus in 15th century Punjab. It is the fifth-largest organised religion in the world, with approximately 30 million Sikhs. Sikhs are expected to embody the qualities of a *Sant-Sipâhî*—a saint-soldier, have control over one's internal vices and be able to be constantly immersed in virtues clarified in the Guru Granth Sahib. The principal beliefs of Sikhi are faith in *Waheguru*—represented

by the phrase *ik ôaEkâr*, meaning one God, who prevails in everything, along with a praxis in which the Sikh is enjoined to engage in social reform through the pursuit of justice for all human beings.

African Traditional

African traditional religion encompasses the traditional religious beliefs of people in Africa. In north Africa, these religions have included traditional Berber religion, ancient Egyptian religion, and Waaq. West African religions include Akan religion, Dahomey (Fon) mythology, Efik mythology, Odinani of the Igbo people, Serer religion, and Yoruba religion, while Bushongo mythology, Mbuti (Pygmy) mythology, Lugbara mythology, Dinka religion, and Lotuko mythology come from central Africa. Southern African traditions include Akamba mythology, Masai mythology, Malagasy mythology, San religion, Lozi mythology, Tumbuka mythology, and Zulu mythology. Bantu mythology is found throughout central, southeast, and southern Africa.

There are also notable African diasporic religions practiced in the Americas, such as Santeria, Candomble, Vodun, Lucumi, Umbanda, and Macumba.

Indigenous and Folk

Indigenous religions or folk religions refers to a broad category of traditional religions that can be characterised by shamanism, animism and ancestor worship, where *traditional* means "indigenous, that which is aboriginal or foundational, handed down from generation to generation...". These are religions that are closely associated with a particular group of people, ethnicity or tribe; they often have no formal creeds or sacred texts. Some faiths are syncretic, fusing diverse religious beliefs and practices.

- Chinese folk religion: the indigenous religions of the Han Chinese, or, by metonymy, of all the populations of the Chinese cultural sphere. It includes Confucianism, Taoism and Wuism, as well as many new religious movements such as Falun Gong and Yiguandao.
- Other folk religions of East Asia and Southeast Asia: Korean shamanism, Japanese Shinto, Satsana Phi, Vietnamese folk religion.
- Australian Aboriginal religions.
- Folk religions of the Americas: Native American religions

Folk religions are often omitted as a category in surveys even in countries where they are widely practiced, e.g. in China.

New

New religious movements include:

- Shinshûkyô is a general category for a wide variety of religious movements founded in Japan since the 19th century. These movements share almost nothing in common except the place of their founding. The largest religious movements centred in Japan include Soka Gakkai, Tenrikyo, and Seicho-No-Ie among hundreds of smaller groups.
- Cao Đài is a syncretistic, monotheistic religion, established in Vietnam in 1926.
- Raëlism is a new religious movement founded in 1974 teaching that humans were created by aliens. It is numerically the world's largest UFO religion.
- Hindu reform movements, such as Ayyavazhi, Swaminarayan Faith and Ananda Marga, are examples of new religious movements within Indian religions.
- Unitarian Universalism is a religion characterized by support for a "free and responsible search for truth and meaning", and has no accepted creed or theology.
- Noahidism is a Biblical-Talmudic and monotheistic ideology for non-Jews based on the Seven Laws of Noah, and on their traditional interpretations within Judaism.
- Scientology teaches that people are immortal beings who have forgotten their true nature. Its method of spiritual rehabilitation is a type of counseling known as auditing, in which practitioners aim to consciously re-experience and understand painful or traumatic events and decisions in their past in order to free themselves of their limiting effects.
- Eckankar is a pantheistic religion with the purpose of making God an everyday reality in one's life.
- Wicca is a neo-pagan religion first popularised in 1954 by British civil servant Gerald Gardner, involving the worship of a God and Goddess.
- Druidry is a religion promoting harmony with nature, and drawing on the practices of the druids.
- Satanism is a broad category of religions that, for example, worship Satan as a deity (Theistic Satanism) or use "Satan" as a symbol of carnality and earthly values (LaVeyan Satanism).

Sociological classifications of religious movements suggest that within any given religious group, a community can resemble various types of structures, including "churches", "denominations", "sects", "cults", and "institutions".

Issues

Economics: While there has been much debate about how religion affects the economy of countries, in general there is a negative correlation between religiosity and the wealth of nations. In other words, the richer a nation is, the less religious it tends to be. However, sociologist and political economist Max Weber has argued that Protestant Christian countries are wealthier because of their Protestant work ethic.

Health

Mayo Clinic researchers examined the association between religious involvement and spirituality, and physical health, mental health, health-related quality of life, and other health outcomes. The authors reported that: "Most studies have shown that religious involvement and spirituality are associated with better health outcomes, including greater longevity, coping skills, and health-related quality of life (even during terminal illness) and less anxiety, depression, and suicide."

The authors of a subsequent study concluded that the influence of religion on health is "largely beneficial", based on a review of related literature. According to academic James W. Jones, several studies have discovered "positive correlations between religious belief and practice and mental and physical health and longevity."

An analysis of data from the 1998 US General Social Survey, whilst broadly confirming that religious activity was associated with better health and well-being, also suggested that the role of different dimensions of spirituality/religiosity in health is rather more complicated. The results suggested "that it may not be appropriate to generalize findings about the relationship between spirituality/religiosity and health from one form of spirituality/religiosity to another, across denominations, or to assume effects are uniform for men and women.

Violence

Charles Selengut characterizes the phrase "religion and violence" as "jarring", asserting that "religion is thought to be opposed to violence and a force for peace and reconciliation. He acknowledges, however, that "the history and scriptures of the world's religions tell stories of violence and war as they speak of peace and love."

Hector Avalos argues that, because religions claim divine favour for themselves, over and against other groups, this sense of righteousness leads to violence because conflicting claims to superiority, based on unverifiable appeals to God, cannot be adjudicated objectively.

Critics of religion Christopher Hitchens and Richard Dawkins go further and argue that religions do tremendous harm to society by using violence to promote their goals, in ways that are endorsed and exploited by their leaders. Regina Schwartz argues that all monotheistic religions are inherently violent because of an exclusivism that inevitably fosters violence against those that are considered outsiders. Lawrence Wechsler asserts that Schwartz isn't just arguing that Abrahamic religions have a violent legacy, but that the legacy is actually genocidal in nature.

Byron Bland asserts that one of the most prominent reasons for the "rise of the secular in Western thought" was the reaction against the religious violence of the 16th and 17th centuries. He asserts that "(t)he secular was a way of living with the religious differences that had produced so much horror. Under secularity, political entities have a warrant to make decisions independent from the need to enforce particular versions of religious orthodoxy. Indeed, they may run counter to certain strongly held beliefs if made in the interest of common welfare. Thus, one of the important goals of the secular is to limit violence."

Nonetheless, believers have used similar arguments when responding to atheists in these discussions, pointing to the widespread imprisonment and mass murder of individuals under atheist states in the twentieth century:

> *And who can deny that Stalin and Mao, not to mention Pol Pot and a host of others, all committed atrocities in the name of a Communist ideology that was explicitly atheistic? Who can dispute that they did their bloody deeds by claiming to be establishing a 'new man' and a religion-free utopia? These were mass murders performed with atheism as a central part of their ideological inspiration, they were not mass murders done by people who simply happened to be atheist. —Dinesh D'Souza*

In response to such a line of argument, however, author Sam Harris writes:

> *The problem with fascism and communism, however, is not that they are too critical of religion; the problem is that they are too much like religions. Such regimes are*

> *dogmatic to the core and generally give rise to personality cults that are indistinguishable from cults of religious hero worship. Auschwitz, the gulag and the killing fields were not examples of what happens when human beings reject religious dogma; they are examples of political, racial and nationalistic dogma run amok. There is no society in human history that ever suffered because its people became too reasonable.* —*Sam Harris*

Richard Dawkins has stated that Stalin's atrocities were influenced not by atheism but by dogmatic Marxism, and concludes that while Stalin and Mao happened to be atheists, they did not do their deeds in the name of atheism. On other occasions, Dawkins has replied to the argument that Adolf Hitler and Josef Stalin were antireligious with the response that Hitler and Stalin also grew moustaches, in an effort to show the argument as fallacious. Instead, Dawkins argues in *The God Delusion* that "What matters is not whether Hitler and Stalin were atheists, but whether atheism systematically influences people to do bad things. There is not the smallest evidence that it does." Dawkins adds that Hitler in fact, repeatedly affirmed a strong belief in Christianity, but that his atrocities were no more attributable to his theism than Stalin's or Mao's were to their atheism. In all three cases, he argues, the perpetrators' level of religiosity was incidental. D'Souza responds that an individual need not explicitly invoke atheism in committing atrocities if it is already implied in his worldview, as is the case in Marxism.

Law

The study of law and religion is a relatively new field, with several thousand scholars involved in law schools, and academic departments including political science, religion, and history since 1980. Scholars in the field are not only focused on strictly legal issues about religious freedom or non-establishment, but also study religions as they are qualified through judicial discourses or legal understanding of religious phenomena. Exponents look at canon law, natural law, and state law, often in a comparative perspective. Specialists have explored themes in western history regarding Christianity and justice and mercy, rule and equity, and discipline and love. Common topics of interest include marriage and the family and human rights. Outside of Christianity, scholars have looked at law and religion links in the Muslim Middle East and pagan Rome.

Studies have focused on secularization. In particular the issue of wearing religious symbols in public, such as headscarves that are banned in French schools, have received scholarly attention in the context of human rights and feminism.

Science

Religious knowledge, according to religious practitioners, may be gained from religious leaders, sacred texts, scriptures, or personal revelation. Some religions view such knowledge as unlimited in scope and suitable to answer any question; others see religious knowledge as playing a more restricted role, often as a complement to knowledge gained through physical observation. Adherents to various religious faiths often maintain that religious knowledge obtained via sacred texts or revelation is absolute and infallible and thereby creates an accompanying religious cosmology, although the proof for such is often tautological and generally limited to the religious texts and revelations that form the foundation of their belief.

In contrast, the scientific method gains knowledge by testing hypotheses to develop theories through elucidation of facts or evaluation by experiments and thus only answers cosmological questions about the universe that can be observed and measured. It develops theories of the world which best fit physically observed evidence. All scientific knowledge is subject to later refinement, or even outright rejection, in the face of additional evidence. Scientific theories that have an overwhelming preponderance of favourable evidence are often treated as *de facto* verities in general parlance, such as the theories of general relativity and natural selection to explain respectively the mechanisms of gravity and evolution.

Regarding religion and science, Albert Einstein states (1940): "For science can only ascertain what is, but not what should be, and outside of its domain value judgments of all kinds remain necessary. Religion, on the other hand, deals only with evaluations of human thought and action; it cannot justifiably speak of facts and relationships between facts...Now, even though the realms of religion and science in themselves are clearly marked off from each other, nevertheless there exist between the two strong reciprocal relationships and dependencies. Though religion may be that which determine the goals, it has, nevertheless, learned from science, in the broadest sense, what means will contribute to the attainment of the goals it has set up."

Related Forms of Thought

Superstition: Superstition has been described as "the incorrect establishment of cause and effect" or a false conception of causation. Religion is more complex and includes social institutions and morality.

But religions may include superstitions or make use of magical thinking. Adherents of one religion sometimes think of other religions as superstition. Some atheists, deists, and skeptics regard religious belief as superstition.

Greek and Roman pagans, who saw their relations with the gods in political and social terms, scorned the man who constantly trembled with fear at the thought of the gods (*deisidaimonia*), as a slave might fear a cruel and capricious master. The Romans called such fear of the gods *superstitio*. Early Christianity was outlawed as a *superstitio Iudaica*, a "Jewish superstition", by Domitian in the 80s AD. In AD 425, when Rome had become Christian, Theodosius II outlawed pagan traditions as superstitious.

The Roman Catholic Church considers superstition to be sinful in the sense that it denotes a lack of trust in the divine providence of God and, as such, is a violation of the first of the Ten Commandments. The Catechism of the Catholic Church states that superstition "in some sense represents a perverse excess of religion" (para. #2110). "Superstition," it says, "is a deviation of religious feeling and of the practices this feeling imposes. It can even affect the worship we offer the true God, e.g., when one attributes an importance in some way magical to certain practices otherwise lawful or necessary. To attribute the efficacy of prayers or of sacramental signs to their mere external performance, apart from the interior dispositions that they demand is to fall into superstition. Cf. Matthew 23:16-22" (para. #2111)

Myth

The word *myth* has several meanings.

1. A traditional story of ostensibly historical events that serves to unfold part of the world view of a people or explain a practice, belief, or natural phenomenon;
2. A person or thing having only an imaginary or unverifiable existence; or
3. A metaphor for the spiritual potentiality in the human being.

Ancient polytheistic religions, such as those of Greece, Rome, and Scandinavia, are usually categorized under the heading of mythology. Religions of pre-industrial peoples, or cultures in development, are similarly called "myths" in the anthropology of religion. The term "myth" can be used pejoratively by both religious and non-religious people. By defining another person's religious stories and beliefs as mythology, one implies that they are less real or true than one's own religious stories and beliefs. Joseph Campbell remarked, "Mythology is often thought of as *other people's* religions, and religion can be defined as mis-interpreted mythology." In sociology, however, the term *myth* has a non-pejorative meaning. There, *myth* is defined as a story that is important for the group whether or not it is objectively or provably true. Examples include the death and resurrection of Jesus, which, to Christians, explains the means by which they are freed from sin and is also ostensibly a historical event. But from a mythological outlook, whether or not the event actually occurred is unimportant. Instead, the symbolism of the death of an old "life" and the start of a new "life" is what is most significant. Religious believers may or may not accept such symbolic interpretations.

Secularism and Irreligion

The terms "atheist" (lack of belief in any gods) and "agnostic" (belief in the unknowability of the existence of gods), though specifically contrary to theistic (e.g. Christian, Jewish, and Muslim) religious teachings, do not by definition mean the opposite of "religious". There are religions (including Buddhism and Taoism), in fact, that classify some of their followers as agnostic, atheistic, or nontheistic. The true opposite of "religious" is the word "irreligious". Irreligion describes an absence of any religion; antireligion describes an active opposition or aversion toward religions in general.

As religion became a more personal matter in Western culture, discussions of society became more focused on political and scientific meaning, and religious attitudes (dominantly Christian) were increasingly seen as irrelevant for the needs of the European world. On the political side, Ludwig Feuerbach recast Christian beliefs in light of humanism, paving the way for Karl Marx's famous characterization of religion as "the opium of the people". Meanwhile, in the scientific community, T.H. Huxley in 1869 coined the term "agnostic," a term—subsequently adopted by such figures as Robert Ingersoll—that, while directly conflicting with and novel to Christian tradition, is accepted and even embraced in some other religions. Later, Bertrand Russell told the world *Why I Am Not a Christian*, which influenced several later authors to discuss their breakaway from their own religious upbringings from Islam to Hinduism.

Some atheists also construct parody religions, for example, the Church of the SubGenius or the Flying Spaghetti Monster, which parodies the equal time argument employed by intelligent design Creationism. Parody religions may also be considered a post-modern approach to religion. For instance, in Discordianism, it may be hard to tell if even these "serious" followers are not just taking part in an even bigger joke. This joke, in turn, may be part of a greater path to enlightenment, and so on *ad infinitum*.

Criticism of Religion

Religious criticism has a long history, going back at least as far as the 5th century BCE. During classical times, there were religious critics in ancient Greece, such as Diagoras "the atheist" of Melos, and in the 1st century BCE in Rome, with Titus Lucretius Carus's *De Rerum Natura*.

During the Middle Ages and continuing into the Renaissance, potential critics of religion were persecuted and largely forced to remain silent. There were notable critics like Giordano Bruno, who was burned at the stake for disagreeing with religious authority.

In the 17th and 18th century with the Enlightenment, thinkers like David Hume and Voltaire criticized religion. In the 19th century, Charles Darwin and the theory of evolution led to increased skepticism about religion. Thomas Huxley, Jeremy Bentham, Karl Marx, Charles Bradlaugh, Robert Ingersol, and Mark Twain were noted 19th-century and early-20th-century critics. In the 20th century, Bertrand Russell, Siegmund Freud, and others continued religious criticism.

Sam Harris, Daniel Dennett, Richard Dawkins, Victor J. Stenger, and the late Christopher Hitchens were active critics during the late 20th century and early 21st century.

Critics consider religion to be outdated, harmful to the individual (e.g. brainwashing of children, faith healing, female genital mutilation, circumcision), harmful to society (e.g. holy wars, terrorism, wasteful distribution of resources), to impede the progress of science, to exert social control, and to encourage immoral acts (e.g. blood sacrifice, discrimination against homosexuals and women, and certain forms of sexual violence such as marital rape). A major criticism of many religions is that they require beliefs that are irrational, unscientific, or unreasonable, because religious beliefs and traditions lack scientific or rational foundations.

Some modern-day critics, such as Bryan Caplan, hold that religion lacks utility in human society; they may regard religion as irrational. Nobel Peace Laureate Shirin Ebadi has spoken out against undemocratic Islamic countries justifying "oppressive acts" in the name of Islam.

Society

A human society is a group of people involved in persistent interpersonal relationships, or a large social grouping sharing the same geographical or social territory, typically subject to the same political authority and dominant cultural expectations. Human societies are characterized by patterns of relationships (social relations) between individuals who share a distinctive culture and institutions; a given society may be described as the sum total of such relationships among its constituent members. In the social sciences, a larger society often evinces stratification or dominance patterns in subgroups..

Insofar as it is collaborative, a society can enable its members to benefit in ways that would not otherwise be possible on an individual basis; both individual and social (common) benefits can thus be distinguished, or in many cases found to overlap.

A society can also consist of like-minded people governed by their own norms and values within a dominant, larger society. This is sometimes referred to as a subculture, a term used extensively within criminology.

More broadly, and especially within structuralist thought, a society may be illustrated as an economic, social, industrial or cultural infrastructure, made up of, yet distinct from, a varied collection of individuals. In this regard society can mean the objective relationships people have with the material world and with other people, rather than "other people" beyond the individual and their familiar social environment.

Members of a society may be from different ethnic groups. A society can be a particular ethnic group, such as the Saxons; a nation state, such as Bhutan; or a broader cultural group, such as a Western society. The word *society* may also refer to an organised voluntary association of people for religious, benevolent, cultural, scientific, political, patriotic, or other purposes. A "society" may even, though more by means of metaphor, refer to a social organism such as an ant colony or any cooperative aggregate such as, for example, in some formulations of artificial intelligence.

Conceptions

Society, in general, addresses the fact that an individual has rather limited means as an autonomous unit. The great apes have always been more (*Bonobo*, *Homo*, *Pan*) or less (*Gorilla*, *Pongo*) social animals, so Robinson Crusoe-like situations are either fictions or unusual corner cases to the ubiquity of social context for humans, who fall between presocial and eusocial in the spectrum of animal ethology.

Human societies are most often organised according to their primary means of subsistence. Social scientists have identified hunter-gatherer societies, nomadic pastoral societies, horticulturalist or simple farming societies, and intensive agricultural societies, also called civilizations. Some consider industrial and post-industrial societies to be qualitatively different from traditional agricultural societies.

Today, anthropologists and many social scientists vigorously oppose the notion of cultural evolution and rigid "stages" such as these. In fact, much anthropological data has suggested that complexity (civilization, population growth and density, specialisation, etc.) does not always take the form of hierarchical social organisation or stratification.

Cultural relativism as a widespread approach or ethic has largely replaced notions of "primitive", better/worse, or "progress" in relation to cultures (including their material culture/technology and social organisation).

According to anthropologist Maurice Godelier, one critical novelty in human society, in contrast to humanity's closest biological relatives (chimpanzees and bonobos), is the parental role assumed by the males, which supposedly would be absent in our nearest relatives for whom paternity is not generally determinable.

In Political Science

Societies may also be structured politically. In order of increasing size and complexity, there are bands, tribes, chiefdoms, and state societies. These structures may have varying degrees of political power, depending on the cultural, geographical, and historical environments that these societies must contend with. Thus, a more isolated society with the same level of technology and culture as other societies is more likely to survive than one in closer proximity to others that may encroach on their resources. A society that is unable to offer an effective response to other societies it competes with will usually be subsumed into the culture of the competing society.

In Sociology

Sociologist Gerhard Lenski differentiates societies based on their level of technology, communication, and economy: (1) hunters and gatherers, (2) simple agricultural, (3) advanced agricultural, (4) industrial, and (5) special (e.g. fishing societies or maritime societies). This is similar to the system earlier developed by anthropologists Morton H. Fried, a conflict theorist, and Elman Service, an integration theorist, who have produced a system of classification for societies in all human cultures based on the evolution of social inequality and the role of the state. This system of classification contains four categories:

- Hunter-gatherer bands (categorization of duties and responsibilities).
- Tribal societies in which there are some limited instances of social rank and prestige.
- Stratified structures led by chieftains.
- Civilizations, with complex social hierarchies and organised, institutional governments.

In addition to this there are:

- Humanity, mankind, upon which rest all the elements of society, including society's beliefs.
- Virtual society, a society based on online identity, which is evolving in the information age.

Over time, some cultures have progressed toward more complex forms of organisation and control. This cultural evolution has a profound effect on patterns of community. Hunter-gatherer tribes settled around seasonal food stocks to become agrarian villages. Villages grew to become towns and cities. Cities turned into city-states and nation-states.

Many societies distribute largess at the behest of some individual or some larger group of people. This type of generosity can be seen in all known cultures; typically, prestige accrues to the generous individual or group. Conversely, members of a society may also shun or scapegoat members of the society who violate its norms. Mechanisms such as gift-giving, joking relationships and scapegoating, which may be seen in various types of human groupings, tend to be institutionalized within a society. Social evolution as a phenomenon carries with it certain elements that could be detrimental to the population it serves.

Some societies bestow status on an individual or group of people when that individual or group performs an admired or desired action. This type of recognition is bestowed in the form of a name, title, manner of dress, or monetary reward. In many societies, adult male or female status is subject to a ritual or process of this type. Altruistic action in the interests of the larger group is seen in virtually all societies. The phenomena of community action, shunning, scapegoating, generosity, shared risk, and reward are common to many forms of society.

Types

Societies are social groups that differ according to subsistence strategies, the ways that humans use technology to provide needs for themselves. Although humans have established many types of societies throughout history, anthropologists tend to classify different societies according to the degree to which different groups within a society have unequal access to advantages such as resources, prestige, or power. Virtually all societies have developed some degree of inequality among their people through the process of social stratification, the division of members of a society into levels with unequal wealth, prestige, or power. Sociologists place societies in three broad categories: pre-industrial, industrial, and postindustrial.

Pre-Industrial

In a pre-industrial society, food production, which is carried out through the use of human and animal labour, is the main economic activity. These societies can be subdivided according to their level of technology and their method of producing food. These subdivisions are hunting and gathering, pastoral, horticultural, agricultural, and feudal.

Hunting and Gathering

Figure: *Starting fire by hand. San people in Botswana.*

The main form of food production in such societies is the daily collection of wild plants and the hunting of wild animals. Hunter-gatherers move around constantly in search of food. As a result, they do not build permanent villages or create a wide variety of artifacts, and usually only form small groups such as bands and tribes. However, some hunting and gathering societies in areas with abundant resources (such as the Tlingit) lived in larger groups and formed complex hierarchical social structures such as chiefdoms. The need for mobility also limits the size of these societies. They generally consist of fewer than 60 people and rarely exceed 100. Statuses within the tribe are relatively equal, and decisions are reached through general agreement. The ties that bind the tribe are more complex than those of the bands. Leadership is personal—charismatic—and used for special purposes only in tribal society. There are no political offices containing real power, and a chief is merely a person of influence, a sort of adviser; therefore, tribal consolidations for collective action are not governmental. The family forms the main social unit, with most societal members being related by birth or marriage. This type of organisation requires the family to carry out most social functions, including production and education.

Pastoral

Pastoralism is a slightly more efficient form of subsistence. Rather than searching for food on a daily basis, members of a pastoral society rely on domesticated herd animals to meet their food needs. Pastoralists live a nomadic life, moving their herds from one pasture to another. Because their food supply is far more reliable, pastoral societies can support larger populations. Since there are food surpluses, fewer people are needed to produce food. As a result, the division of labour (the specialisation by individuals or groups in the performance of specific economic activities) becomes more complex. For example, some people become craftworkers, producing tools, weapons, and jewellery. The production of goods encourages trade. This trade helps to create inequality, as some families acquire more goods than others do. These families often gain power through their increased wealth. The passing on of property from one generation to another helps to centralize wealth and power. Over time emerge hereditary chieftainships, the typical form of government in pastoral societies.

Horticultural

Fruits and vegetables grown in garden plots that have been cleared from the jungle or forest provide the main source of food in a horticultural society. These societies have a level of technology and complexity similar to pastoral societies. Some horticultural groups use the slash-and-burn method to raise crops. The wild vegetation is cut and burned, and ashes are used as fertilizers. Horticulturists use human labour and simple tools to cultivate the land for one or more seasons. When the land becomes barren, horticulturists clear a new plot and leave the old plot to revert to its natural state. They may return to the original land several years later and begin the process again. By rotating their garden plots, horticulturists can stay in one area for a fairly long period of time. This allows them to build semipermanent or permanent villages. The size of a village's population depends on the amount of land available for farming; thus villages can range from as few as 30 people to as many as 2000.

As with pastoral societies, surplus food leads to a more complex division of labour. Specialised roles in horticultural societies include craftspeople, shamans (religious leaders), and traders. This role specialisation allows people to create a wide variety of artifacts. As in pastoral societies, surplus food can lead to inequalities in wealth and power within horticultural political systems, developed because of the settled nature of horticultural life.

Agrarian

Agrarian societies use agricultural technological advances to cultivate crops over a large area. Sociologists use the phrase Agricultural Revolution to refer to the technological changes that occurred as long as 8,500 years ago that led to cultivating crops and raising farm animals. Increases in food supplies then led to larger populations than in earlier communities. This meant a greater surplus, which resulted in towns that became centres of trade supporting various rulers, educators, craftspeople, merchants, and religious leaders who did not have to worry about locating nourishment.

Greater degrees of social stratification appeared in agrarian societies. For example, women previously had higher social status because they shared labour more equally with men. In hunting and gathering societies, women even gathered more food than men. However, as food stores improved and women took on lesser roles in providing food for the family, they increasingly became subordinate to men. As villages and towns expanded into neighbouring areas, conflicts with other communities inevitably occurred. Farmers provided warriors with food in exchange for protection against invasion by enemies. A system of rulers with high social status also appeared. This nobility organised warriors to protect the society from invasion. In this way, the nobility managed to extract goods from "lesser" members of society.

Feudal

Feudalism was a form of society based on ownership of land. Unlike today's farmers, vassals under feudalism were bound to cultivating their lord's land. In exchange for military protection, the lords exploited the peasants into providing food, crops, crafts, homage, and other services to the landowner. The estates of the realm system of feudalism was often multigenerational; the families of peasants may have cultivated their lord's land for generations.

Industrial

Between the 15th and 16th centuries, a new economic system emerged that began to replace feudalism. Capitalism is marked by open competition in a free market, in which the means of production are privately owned. Europe's exploration of the Americas served as one impetus for the development of capitalism. The introduction of foreign metals, silks, and spices stimulated great commercial activity in European societies. Industrial societies rely heavily on machines powered by fuels for the production of goods. This produced further dramatic increases in efficiency. The increased efficiency of production of the industrial revolution produced an even greater surplus than before. Now the surplus was not just agricultural goods, but also manufactured goods. This larger surplus caused all of the changes discussed earlier in the domestication revolution to become even more pronounced.

Once again, the population boomed. Increased productivity made more goods available to everyone. However, inequality became even greater than before. The breakup of agricultural-based feudal societies caused many people to leave the land and seek employment in cities. This created a great surplus of labour and gave capitalists plenty of laborers who could be hired for extremely low wages.

Post-Industrial

Postindustrial societies are societies dominated by information, services, and high technology more than the production of goods. Advanced industrial societies are now seeing a shift toward an increase in service sectors over manufacturing and production. The U.S. is the first country to have over half of its work force employed in service industries. Service industries include government, research, education, health, sales, law, banking, and so on. It is still too early to identify and understand all the ramifications this new kind of society will have for social life. In fact, even the phrase "postindustrial" belies the fact that we don't yet quite know what will follow industrial societies or the forms they will take.

Contemporary Usage

The term "society" is currently used to cover both a number of political and scientific connotations as well as a variety of associations.

Western

The development of the Western world has brought with it the emerging concepts of Western culture, politics, and ideas, often referred to simply as "Western society". Geographically, it covers at the very least the countries of Western Europe, North America, Australia, and New Zealand. It sometimes also includes Eastern Europe, South America, and Israel.

The cultures and lifestyles of all of these stem from Western Europe. They all enjoy relatively strong economies and stable governments, allow freedom of religion, have chosen democracy as a form of governance, favour capitalism and international trade, are heavily influenced by Judeo-Christian values, and have some form of political and military alliance or cooperation.

Information

Although the concept of information society has been under discussion since the 1930s, in the modern world it is almost always applied to the manner in which information technologies have impacted society and culture. It therefore covers the effects of computers and telecommunications on the home, the workplace, schools, government, and various communities and organisations, as well as the emergence of new social forms in cyberspace.

One of the European Union's areas of interest is the information society. Here policies are directed towards promoting an open and competitive digital economy, research into information and communication technologies, as well as their application to improve social inclusion, public services, and quality of life.

The International Telecommunications Union's World Summit on the Information Society in Geneva and Tunis (2003 and 2005) has led to a number of policy and application areas where action is required. These include:

- promotion of ICTs for development;
- information and communication infrastructure;
- access to information and knowledge;
- capacity building;
- building confidence and security in the use of ICTs;
- enabling environment;
- ICT applications in the areas of government, business, learning, health, employment, environment, agriculture and science;
- cultural and linguistic diversity and local content;
- media;
- ethical dimensions of the information society; and
- international and regional cooperation.

Knowledge

As access to electronic information resources increased at the beginning of the 21st century, special attention was extended from the information society to the knowledge society. An analysis by the Irish government stated, "The capacity to manipulate, store and transmit large quantities of information cheaply has increased at a staggering rate over recent years. The digitisation of information and the associated pervasiveness of the Internet are facilitating a new intensity in the application of knowledge to economic activity, to the extent that it has become the predominant factor in the creation of wealth. As much as 70 to 80 percent of economic growth is now said to be due to new and better knowledge."

The Second World Summit on the Knowledge Society, held in Chania, Crete, in September 2009, gave special attention to the following topics:

- business and enterprise computing;
- technology-enhanced learning;
- social and humanistic computing;
- culture, tourism and technology;
- e-government and e-democracy;
- innovation, sustainable development, and strategic management;
- service science, management, and engineering;
- intellectual and human capital development;
- ICTs for ecology and the green economy;
- future prospects for the knowledge society; and
- technologies and business models for the creative industries.

Other Uses

People of many nations united by common political and cultural traditions, beliefs, or values are sometimes also said to form a society (such as Judeo-Christian, Eastern, and Western). When used in this context, the term is employed as a means of contrasting two or more "societies" whose members represent alternative conflicting and competing worldviews.

Some academic, professional, and scientific associations describe themselves as *societies* (for example, the American Mathematical Society, the American Society of Civil Engineers, or the Royal Society).

In some countries, e.g. the United States, France, and Latin America, the term "society' is used in commerce to denote a partnership between investors or the start of a business. In the United Kingdom, partnerships are not called societies, but co-operatives or mutuals are often known as societies (such as friendly societies and building societies).

Chapter 3

Social Class

Social class (or simply "class"), as in a class society, is a set of concepts in the social sciences and political theory centred on models of social stratification in which people are grouped into a set of hierarchical social categories, the most common being the upper, middle, and lower classes.

Class is an essential object of analysis for sociologists, political scientists, anthropologists, and social historians. However, there is not a consensus on the best definition of the term "class," and the term has different contextual meanings. In common parlance, the term "social class" is usually synonymous with "socio-economic class," defined as "people having the same social, economic, or educational status" e.g., "the working class"; "an emerging professional class." However, academics distinguish social class and socioeconomic status, with the former referring to one's relatively stable sociocultural background and the latter referring to one's current social and economic situation and, consequently, being more changeable over time.

The precise measurements of what determines social class in society has varied over time. According to philosopher Karl Marx, "class" is determined entirely by one's relationship to the means of production, the classes in modern capitalist society being the "proletarians": those who work but do not own the means of production, the "bourgeoisie": those who invest and live off of the surplus generated by the former, and the aristocracy that has land as a means of production.

The term "class" is etymologically derived from the Latin *classis*, which was used by census takers to categorize citizens by wealth, in order to determine military service obligations.

In the late 18th century, the term "class" began to replace classifications such as estates, rank, and orders as the primary means of organising society into hierarchical divisions. This corresponded to a general decrease in significance ascribed to hereditary characteristics, and increase in the significance of wealth and income as indicators of position in the social hierarchy.

History

Historically social class and behaviour was sometimes laid down in law. For example, permitted mode of dress in some times and places was strictly regulated, with sumptuous dressing only for the high ranks of society and aristocracy; sumptuary laws stipulated the dress and jewellery appropriate for a person's social rank and station.

Theoretical Models

Definitions of social classes reflect a number of sociological perspectives, informed by anthropology, economics, psychology, and sociology. The major perspectives historically have been Marxism and Structural functionalism. The common *stratum model of class* divides society into a simple hierarchy of working class, middle class and upper class. Within academia, two broad schools of definitions emerge: those aligned with 20th-century sociological stratum models of class society, and those aligned with the 19th-century historical materialist economic models of the Marxists and anarchists.

Another distinction can be drawn between *analytical* concepts of social class, such as the *Marxist* and *Weberian* traditions, and the more *empirical* traditions such as socio-economic status approach, which notes the correlation of income, education and wealth with social outcomes without necessarily implying a particular theory of social structure.

Marxist

> *"Classes are large groups of people differing from each other by the place they occupy in a historically determined system of social production, by their relation (in most cases fixed and formulated in law) to the means of production, by their role in the social organisation of labour, and, consequently, by the dimensions of the share of social wealth of which they dispose and the mode of acquiring it."* *Vladimir Lenin,* A Great Beginning - *June, 1919*

For Marx, class is a combination of objective and subjective factors. Objectively, a class shares a common relationship to the means of production. Subjectively, the members will necessarily have some perception ("class consciousness") of their similarity and common interest. Class consciousness is not simply an awareness of one's own class interest but is also a set of shared views regarding how society should be organised legally, culturally, socially and politically. These class relations are reproduced through time.

In Marxist theory, the class structure of the capitalist mode of production is characterized by the conflict between two main classes: the bourgeoisie, the capitalists who own the means of production, and the much larger proletariat (or 'working class') who must sell their own labour power. This is the fundamental economic structure of work and property, a state of inequality that is normalized and reproduced through cultural ideology.

Marxists explain the history of "civilized" societies in terms of a war of classes between those who control production and those who produce the goods or services in society. In the Marxist view of capitalism, this is a conflict between capitalists (bourgeoisie) and wage-workers (the proletariat). For Marxists, class antagonism is rooted in the situation that control over social production necessarily entails control over the class which produces goods—in capitalism this is the exploitation of workers by the bourgeoisie.

Furthermore, "in countries where modern civilization has become fully developed, a new class of petty bourgeois has been formed". "An industrial army of workmen, under the command of a capitalist, requires, like a real army, officers (managers) and sergeants (foremen, over-lookers) who, while the work is being done, command in the name of the capitalist".

Marx himself argued that it was the goal of the proletariat itself to displace the capitalist system with socialism, changing the social relationships underpinning the class system and then developing into a future communist society in which: "..the free development of each is the condition for the free development of all." (Communist Manifesto) This would mark the beginning of a classless society in which human needs rather than profit would be motive for production. In a society with democratic control and production for use, there would be no class, no state and no need for money.

Weberian

Max Weber formulated a three-component theory of stratification, that saw political power as an interplay between "class", "status" and "group power". Weber believed that class position was determined by a person's skills and education, rather than by their relationship to the means of production. Both Marx and Weber agreed that social stratification was undesirable, however where Marx believed that stratification would only disappear along with capitalism and private property, Weber believed that the solution lay in providing "equal opportunity" within a competitive, capitalist system.

Weber derived many of his key concepts on social stratification by examining the social structure of Germany. He noted that contrary to Marx's theories, stratification was based on more than simply ownership of capital. Weber examined how many members of the aristocracy lacked economic wealth yet had strong political power. Many wealthy families lacked prestige and power, for example, because they were Jewish. Weber introduced three independent factors that form his theory of stratification hierarchy; class, status, and power:

- Class: A person's economic position in a society. Weber differs from Marx in that he does not see this as the supreme factor in stratification. Weber noted how managers of corporations or industries control firms they do not own.
- Status: A person's prestige, social honour, or popularity in a society. Weber noted that political power was not rooted in capital value solely, but also in one's individual status. Poets or saints, for example, can possess immense influence on society with often little economic worth.
- Power: A person's ability to get their way despite the resistance of others. For example, individuals in state jobs, such as an employee of the Federal Bureau of Investigation, or a member of the United States Congress, may hold little property or status but they still hold immense power.

Great British Class Survey

On April 2, 2013 the results of a survey conducted by the BBC developed in collaboration with academic experts and slated to be published in the journal *Sociology* were published online. The results released were based on a survey of 160,000 residents of the United Kingdom most of whom lived in England and described themselves as "white." Class was defined and measured according to the amount and kind of economic, cultural, and social resources reported. Economic capital was defined as income and assets; cultural capital as amount and type of cultural interests and activities, and social capital as the quantity and social status of their friends, family and personal and business contacts. This theoretical framework was developed by Pierre Bourdieu who first published his theory of social distinction in 1979.

The Common Three-Stratum Model

Today, concepts of social class often assume three general categories: a very wealthy and powerful *upper class* that owns and controls the means of production; a *middle class* of professional workers, small business owners, and low-level managers; and a *lower class*, who rely on low-paying wage jobs for their livelihood and often experience poverty.

Upper Class

The upper class is the social class composed of those who are rich, well-born, or both. They usually wield the greatest political power. In some countries, wealth alone is sufficient to allow entry into the upper class. In others, only people who are born or marry into certain aristocratic bloodlines are considered members of the upper class, and those who gain great wealth through commercial activity are looked down upon by the "old rich" as *nouveau riche*. In the United Kingdom, for example, the upper classes are the aristocracy and royalty, with wealth playing a less important role in class status. Many aristocratic peerages or titles have 'seats' attached to them, with the holder of the title (e.g. Earl of Bristol) and his family being the custodians of the house, but not the owners. Many of these require high expenditures, so wealth is typically needed. Many aristocratic peerages and their homes are parts of estates, owned and run by the title holder with moneys generated by the land, rents, or other sources wealth. In America, however, where there is no aristocracy or royalty, the upper class status belongs to the extremely wealthy, the so-called 'super-rich', though there is some tendency even in America for those with old family wealth to look down on those who have earned their money in business, the struggle between New Money and Old Money.

The upper class is generally contained within the richest one or two percent of the population. Members of the upper class are often born into it, and are distinguished by immense wealth which is passed from generation to generation in the form of estates. Sometimes members of the upper class are called "the one percent".

Middle Class

The middle class is the most contested of the three categories, the broad group of people in contemporary society who fall socio-economically between the lower and upper classes. One example of the contest of this term is that in the United States "middle class" is applied very broadly and includes people who would elsewhere be considered working class. Middle class workers are sometimes called "white-collar workers". Theorists such as Ralf Dahrendorf have noted the tendency toward an enlarged middle class in modern Western societies, particularly in relation to the necessity of an educated work force in technological economies. Perspectives concerning globalization and neocolonialism, such as dependency theory, suggest this is due to the shift of low-level labour to developing nations and the Third World.

Lower Class

Figure: *In the United States the lowest stratum of the working class, the underclass, often lives in urban areas with low-quality civil services.*

Lower class (occasionally described as working class) are those employed in low-paying wage jobs with very little economic security. The term "lower class" also refers to persons with low income.

The working class is sometimes separated into those who are employed but lacking financial security, and an underclass—those who are long-term unemployed and/or homeless, especially those receiving welfare from the state. The latter is analogous to the Marxist term *"lumpenproletariat"*. Members of the working class are sometimes called blue-collar workers. In the United States, the terms working class and blue-collar may refer to employed and hard-working members of the middle-middle and lower-middle class, while the upper-middle class in the United States often refers to employment positions that require a college or graduate degree.

Consequences of Class Position

A person's socioeconomic class has wide-ranging effects. It may determine the schools they are able to attend, the jobs open to them, who they may marry, and their treatment by police and the courts.

Education

A person's social class has a significant impact on their educational opportunities. Not only are upper-class parents able to send their children to exclusive schools that are perceived to be better, but in many places state-supported schools for children of the upper class are of a much higher quality than those the state provides for children of the lower classes. This lack of good schools is one factor that perpetuates the class divide across generations.

In 1977, British cultural theorist Paul Willis published a study titled "Learning to Labour", in which he investigated the connection between social class and education. In his study, he found that a group of working class schoolchildren had developed an antipathy towards the acquisition of knowledge as being outside their class, and therefore undesirable, perpetuating their presence in the working class.

Health and Nutrition

A person's social class has a significant impact on their physical health, their ability to receive adequate medical care and nutrition, and their life expectancy.

Lower-class people experience a wide array of health problems as a result of their economic status. They are unable to use health care as often, and when they do it is of lower quality, even though they generally tend to experience a much higher rate of health issues. Lower-class families have higher rates of infant mortality, cancer, cardiovascular disease, and disabling physical injuries. Additionally, poor people tend to work in much more hazardous conditions, yet generally have much less (if any) health insurance provided for them, as compared to middle and upper class workers.

Employment

The conditions at a person's job vary greatly depending on class. Those in the upper-middle class and middle class enjoy greater freedoms in their occupations. They are usually more respected, enjoy more diversity, and are able to exhibit some authority. Those in lower classes tend to feel more alienated and have lower work satisfaction overall. The physical conditions of the workplace differ greatly between classes. While middle-class workers may "suffer alienating conditions" or "lack of job satisfaction", blue-collar workers are more apt to suffer alienating, often routine, work with obvious physical health hazards, injury, and even death.

Class Conflict

Class conflict, frequently referred to as "class warfare" or "class struggle," is the tension or antagonism which exists in society due to competing socioeconomic interests and desires between people of different classes.

For Marx, the history of class society was a history of class conflict. He pointed to the successful rise of the bourgeoisie, and the necessity of revolutionary violence—a heightened form of class conflict—in securing the bourgeoisie rights that supported the capitalist economy.

Marx believed that the exploitation and poverty inherent in capitalism were a pre-existing form of class conflict. Marx believed that wage laborers would need to revolt to bring about a more equitable distribution of wealth and political power.

Classless Society

"Classless society" refers to a society in which no one is born into a social class. Distinctions of wealth, income, education, culture, or social network might arise and would only be determined by individual experience and achievement in such a society.

Since these distinctions are difficult to avoid, advocates, such as Anarchists, communists, etc. of a classless society propose various means to achieve and maintain it and attach varying degrees of importance to it as an end in their overall programs/philosophy.

Relationship between Ethnicity and Class

Race and other large-scale groupings can also influence class standing. The association of particular ethnic groups with class statuses is common in many societies. As a result of conquest or internal ethnic differentiation, a ruling class is often ethnically homogenous and particular races or ethnic groups in some societies are legally or customarily restricted to occupying particular class positions. Which ethnicities are considered as belonging to high or low classes varies from society to society. In modern societies strict legal links between ethnicity and class have been drawn, such as in apartheid, the caste system in Africa, the position of the Burakumin in Japanese society, and the Casta system in Latin America.

Economic Inequality

Economic inequality (also described as the gap between rich and poor, income inequality, wealth disparity, wealth and income differences or wealth gap) is the state of affairs in which assets, wealth, or income are distributed unequally among individuals in a group, among groups in a population, or among countries. The issue of economic inequality can implicate notions of equity, equality of outcome, and equality of opportunity.

Opinions differ on the importance of economic inequality and its effects. Some studies have emphasized inequality as a growing social problem. While some inequality may promote investment, too much inequality may be destructive. Income inequality can hinder long term growth, but can also help long term growth. Statistical studies comparing inequality to year-over-year economic growth have been inconclusive; however in 2011, researchers from the International Monetary Fund published work which indicated that income equality increased the duration of countries' economic growth spells more than free trade, low government corruption, foreign investment, or low foreign debt.

Economic inequality varies between societies, historical periods, economic structures and systems. The term can refer to cross sectional distribution of income or wealth at any particular period, or to the lifetime income and wealth over longer periods of time. There are various numerical indices for measuring economic inequality. A widely used one is the Gini coefficient, but there are also many other methods.

Extent

A study entitled "Divided we Stand: Why Inequality Keeps Rising" by the Organisation for Economic Co-operation and Development (OECD) reported its conclusions on the causes, consequences and policy implications for the ongoing intensification of the extremes of wealth and poverty across its 22 member nations (OECD 2011-12-05).

- "Income inequality in OECD countries is at its highest level for the past half century. The average income of the richest 10% of the population is about nine times that of the poorest 10% across the OECD, up from seven times 25 years ago."
- Wealth inequality in the United States has increased further from already high levels.
- Referring to median incomes for the upper 10% contrasted with medians for the lowest 10%, "Other traditionally more egalitarian countries, such as Germany, Denmark and Sweden, have seen the gap between rich and poor expand from 5 to 1 in the 1980s, to 6 to 1 today."

A study by the World Institute for Development Economics Research at United Nations University reports that the richest 1% of adults alone owned 40% of global assets in the year 2000. The *three* richest people in the world possess more financial assets than the lowest 48 nations combined. The combined wealth of the "10 million dollar millionaires" grew to nearly $41 trillion in 2008. A January 2014 report by Oxfam claims that the 85 wealthiest individuals in the world have a combined wealth equal to that of the bottom 50% of the world's population, or about 3.5 billion people. According to a *Los Angeles Times* analysis of the report, the wealthiest 1% owns 46% of the world's wealth; the 85 richest people, a small part of the wealthiest 1%, own about 0.7% of the human population's wealth, which is the same as the bottom half of the population. An October 2014 study by Credit Suisse also claims that the top 1% now own nearly half of the world's wealth and that the accelerating disparity could trigger a recession.

According to PolitiFact the top 400 richest Americans "have more wealth than half of all Americans combined." According to the New York Times on July 22, 2014, the "richest 1 percent in the United States now own more wealth than the bottom 90 percent". Inherited wealth may help explain why many Americans who have become rich may have had a "substantial head start". In September 2012, according to the Institute for Policy Studies, "over 60 percent" of the Forbes richest 400 Americans "grew up in substantial privilege".

The existing data and estimates suggest a large increase in international (and more generally inter-macroregional) component between 1820 and 1960. It might have slightly decreased since that time at the expense of increasing inequality within countries.

The United Nations Development Programme in 2014 asserted that greater investments in social security, jobs and laws that protect vulnerable populations are necessary to prevent widening income inequality....

There is a significant difference in the measured wealth distribution and the public's understanding of wealth distribution. Michael Norton of the Harvard Business School and Dan Ariely of the Department of Psychology at Duke University found this to be true in their research, done in 2011. The actual wealth going to the top quintile in 2011 was around 84% where as the average amount of wealth that the general public estimated to go to the top quintile was around 58%.

Causes

There are many reasons for economic inequality within societies. Recent growth in overall income inequality, at least within the OECD countries, has been driven mostly by increasing inequality in wages and salaries. Economist Thomas Piketty, who specialises in the study of economic inequality, argues that widening economic disparity is an inevitable phenomenon of free market capitalism when the rate of return of capital (r) is greater than the rate of growth of the economy (g).

Common factors thought to impact economic inequality include:

- labour market outcomes
- globalization
- technological changes
- policy reforms
- more regressive taxation
- plutocracy
- computerization and increased technology
- ethnic discrimination
- gender discrimination
- nepotism
- variation in natural ability
- neoliberalism

Theoretical Frameworks

Neoclassical Economics: Neoclassical economics views inequalities in the distribution of income as arising from differences in productivity, and attribute rising inequality to rising differences in the productivity of different groups of workers. In this perspective, wages and profits are determined by the marginal productivity of each individual in the economy. Thus rising inequalities are merely a reflection of the productivity gap between highly-paid professions and lower-paid professions.

Marxian Economics

In Marxian economic analysis, rising income inequality is an inherent feature of capitalism. In this analysis, capitalist firms increasingly substitute workers for capital equipment under competitive pressures to reduce costs and maximize profit. Over the long-term, this trend increases the organic composition of capital, meaning that less labour inputs (workers) are required in proportion to capital inputs, increasing unemployment and the size of the reserve army of labour. This process exerts a downward pressure on wages. The substitution of labour for capital equipment (job automation) increases productivity per worker and thus profits for the capitalist class, resulting in a situation of relatively stagnant wages for the working class amidst rising levels of property income for the capitalist class. Therefore, Marxian economics attributes rising inequality to both the ownership structure of capitalist economies and to rising job automation conflicting with the requirements of the wage labour system.

Labour Market

A major cause of economic inequality within modern market economies is the determination of wages by the market. Some small part of economic inequality is caused by the differences in the supply and demand for different types of work. However, where competition is imperfect; information unevenly distributed; opportunities to acquire education and skills unequal; and since many such imperfect conditions exist in virtually every market, there is in fact little presumption that markets are in general efficient. This means that there is an enormous potential role for government to correct these market failures.

In a purely capitalist mode of production (i.e. where professional and labour organisations cannot limit the number of workers) the workers wages will not be controlled by these organisations, or by the employer, but rather by the market. Wages work in the same way as prices for any other good. Thus, wages can be considered as a function of market price of skill. And therefore, inequality is driven by this price. Under the law of supply and demand, the price of skill is determined by a race between the demand for the skilled worker and the supply of the skilled worker. "On the other hand, markets can also concentrate wealth, pass environmental costs on to society, and abuse workers and consumers." "Markets, by themselves, even when they are stable, often lead to high levels of inequality, outcomes that are widely viewed as unfair." Employers who offer a below market wage will find that their business is chronically understaffed. Their competitors will take advantage of the situation by offering a higher wage to snatch up the best of their labour. For a businessman who has the profit motive as the prime interest, it is a losing proposition to offer below or above market wages to workers.

A job where there are many workers willing to work a large amount of time (high supply) competing for a job that few require (low demand) will result in a low wage for that job. This is because competition between workers drives down the wage. An example of this would be jobs such as dish-washing or customer service. Competition amongst workers tends to drive down wages due to the expendable nature of the worker in relation to his or her particular job. A job where there are few able or willing workers (low supply), but a large need for the positions (high demand), will result in high wages for that job. This is because competition between employers *for employees* will drive up the wage. Examples of this would include jobs that require highly developed skills, rare abilities, or a high level of risk. Competition amongst employers tends to drive up wages due to the nature of the job, since there is a relative shortage of workers for the particular position. Professional and labour organisations may limit the supply of workers which results in higher demand and greater incomes for members. Members may also receive higher wages through collective bargaining, political influence, or corruption.

These supply and demand interactions result in a gradation of wage levels within society that significantly influence economic inequality. Polarization of wages does not explain the accumulation of wealth and very high incomes among the 1%. Joseph Stiglitz believes that "It is plain that markets must be tamed and tempered to make sure they work to the benefit of most citizens."

Taxes

Another cause is the rate at which income is taxed coupled with the progressivity of the tax system. A progressive tax is a tax by which the tax rate increases as the taxable base amount increases. In a progressive tax system, the level of the top tax rate will often have a direct impact on the level of inequality within a society, either increasing it or decreasing it, provided that income does not change as a result of the change in tax regime. Additionally, steeper tax progressivity applied to social spending can result in a more equal distribution of income across the board. The difference between the Gini index for an income distribution before taxation and the Gini index after taxation is an indicator for the effects of such taxation.

There is debate between politicians and economists over the role of tax policy in mitigating or exacerbating wealth inequality. Economists such as Paul Krugman, Peter Orszag, and Emmanuel Saez have argued that tax policy in the post World War II era has indeed increased income inequality by enabling the wealthiest Americans far greater access to capital than lower-income ones.

Education

An important factor in the creation of inequality is variation in individuals' access to education. Education, especially in an area where there is a high demand for workers, creates high wages for those with this education, however, increases in education first increase and then decrease growth as well as income inequality. As a result, those who are unable to afford an education, or choose not to pursue optional education, generally receive much lower wages. The justification for this is that a lack of education leads directly to lower incomes, and thus lower aggregate savings and investment. In particular, the increase in family income and wealth inequality leads to greater dispersion of educational attainment, primarily because those at the bottom of the educational distribution have fallen further below the average level of education. Conversely, education raises incomes and promotes growth because it helps to unleash the productive potential of the poor.

In 2014, economists with the Standard & Poor's rating agency concluded that the widening disparity between the U.S.'s wealthiest citizens and the rest of the nation had slowed its recovery from the 2008-2009 recession and made it more prone to boom-and-bust cycles. To partially remedy the wealth gap and the resulting slow growth, S&P recommended increasing access to education. It estimated that if the average United States worker had completed just one more year of school, it would add an additional $105 billion in growth to the country's economy over five years.

During the mass high school education movement from 1910–1940, there was an increase in skilled workers, which led to a decrease in the price of skilled labour. High school education during the period was designed to equip students with necessary skill sets to be able to perform at work. In fact, it differs from the present high school education, which is regarded as a stepping-stone to acquire college and advanced degrees. This decrease in wages caused a period of compression and decreased inequality between skilled and unskilled workers. Education is very important for the growth of the economy, however educational inequality in gender also influence towards the economy. Lagerlof and Galor stated that gender inequality in education can result to low economic growth, and continued gender inequality in education, thus creating a poverty trap. It is suggested that a large gap in male and female education may indicate backwardness and so may be associated with lower economic growth, which can explain why there is economic inequality between countries.

More of Barro studies also find that female secondary education is positively associated with growth. His findings show that countries with low female education; increasing it has little effect on economic growth, however in countries with high female education, increasing it significantly boosts economic growth. More and better education is a prerequisite for rapid economic development around the world. Education stimulates economic growth and improves people's lives through many channels.

By increasing the efficiency of the labour force it create better conditions for good governance, improving health and enhancing equality. Labour market success is linked to schooling achievement, the consequences of widening disparities in schooling is likely to be further increases in earnings inequality

Deregulation and Decline of Unions

John Schmitt and Ben Zipperer (2006) of the CEPR point to economic liberalism and the reduction of business regulation along with the decline of union membership as one of the causes of economic inequality. In an analysis of the effects of intensive Anglo-American neoliberal policies in comparison to continental European neoliberalism, where unions have remained strong, they concluded "The U.S. economic and social model is associated with substantial levels of social exclusion, including high levels of income inequality, high relative and absolute poverty rates, poor and unequal educational outcomes, poor health outcomes, and high rates of crime and incarceration. At the same time, the available evidence provides little support for the view that U.S.-style labour-market flexibility dramatically improves labour-market outcomes. Despite popular prejudices to the contrary, the U.S. economy consistently affords a lower level of economic mobility than all the continental European countries for which data is available."

Globalization

Trade liberalization may shift economic inequality from a global to a domestic scale. When rich countries trade with poor countries, the low-skilled workers in the rich countries may see reduced wages as a result of the competition, while low-skilled workers in the poor countries may see increased wages. Trade economist Paul Krugman estimates that trade liberalisation has had a measurable effect on the rising inequality in the United States. He attributes this trend to increased trade with poor countries and the fragmentation of the means of production, resulting in low skilled jobs becoming more tradeable. However, he concedes that the effect of trade on inequality in America is minor when compared to other causes, such as technological innovation, a view shared by other experts. Lawrence Katz estimates that trade has only accounted for 5-15% of rising income inequality. Robert Lawrence argues that technological innovation and automation has meant that low-skilled jobs have been replaced by machine labour in wealthier nations, and that wealthier countries no longer have significant numbers of low-skilled manufacturing workers that could be affected by competition from poor countries.

Gender

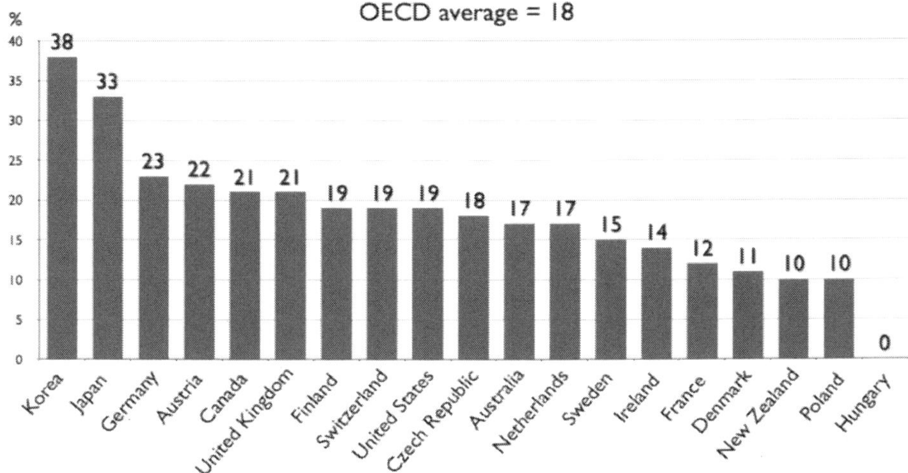

Figure: *The gender gap in median earnings of full-time employees according to the OECD 2008.*

In many countries, there is a gender income gap which favours males in the labour market. For example, the median full-time salary for U.S. women is 77% of that of U.S. men. Several factors other than discrimination may contribute to this gap. On average, women are more likely than men to consider factors other than pay when looking for work, and may be less willing to travel or relocate. Thomas Sowell, in his book Knowledge and Decisions, claims that this difference is due to women not taking jobs due to marriage or pregnancy, but income studies show that does not explain the entire difference. A U.S. Census's report stated that in US once other factors are accounted for there is still a difference in earnings between women and men. The income gap in other countries ranges from 53% in Botswana to -40% in Bahrain.

Gender inequality and discrimination is argued to cause and perpetuate poverty and vulnerability in society as a whole. Gender Equity Indices seek to provide the tools to demonstrate this feature of equity.

19th century socialists like Robert Owen, William Thompson, Anna Wheeler and August Bebel argued that the economic inequality between genders was the leading cause of economic inequality; however Karl Marx and Fredrick Engels believed that the inequality between social classes was the larger cause of inequality.

Economic Development

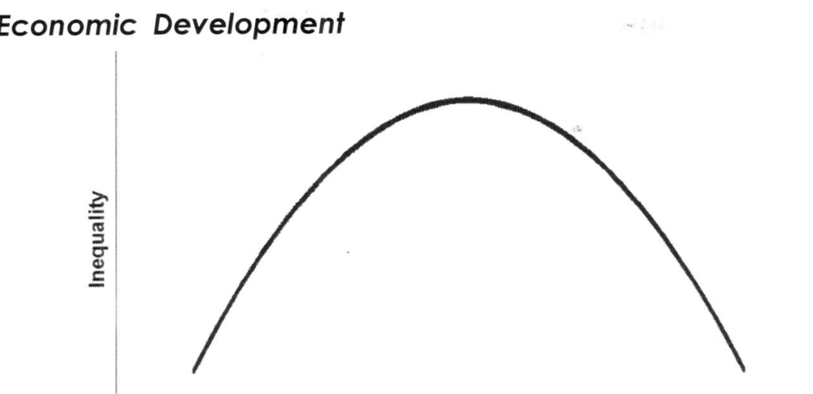

Figure: *A Kuznets curve*

Economist Simon Kuznets argued that levels of economic inequality are in large part the result of stages of development. According to Kuznets, countries with low levels of development have relatively equal distributions of wealth. As a country develops, it acquires more capital, which leads to the owners of this capital having more wealth and income and introducing inequality. Eventually, through various possible redistribution mechanisms such as social welfare programs, more developed countries move back to lower levels of inequality.

Plotting the relationship between level of income and inequality, Kuznets saw middle-income developing economies level of inequality bulging out to form what is now known as the Kuznets curve. Kuznets demonstrated this relationship using cross-sectional data. However, more recent testing of this theory with superior panel data has shown it to be very weak. Kuznets' curve predicts that income inequality will eventually decrease given time. As an example, income inequality did fall in the United States during its High school movement from 1910 to 1940 and thereafter. However, recent data shows that the level of income inequality began to rise after the 1970s. This does not necessarily disprove Kuznets' theory. It may be possible that another Kuznets' cycle is occurring, specifically the move from the manufacturing sector to the service sector. This implies that it may be possible for multiple Kuznets' cycles to be in effect at any given time.

Individual Preferences

Related to cultural issues, diversity of preferences within a society may contribute to economic inequality. When faced with the choice between working harder to earn more money or enjoying more leisure time, equally capable individuals with identical earning potential may choose different strategies. The trade-off between work and leisure is particularly important in the supply side of the labour market in labour economics.

Likewise, individuals in a society often have different levels of risk aversion. When equally-able individuals undertake risky activities with the potential of large payoffs, such as starting new businesses, some ventures succeed and some fail. The presence of both successful and unsuccessful ventures in a society results in economic inequality even when all individuals are identical.

Wealth Concentration

Wealth concentration is a theoretical process by which, under certain conditions, newly created wealth concentrates in the possession of already-wealthy individuals or entities. According to this theory, those who already hold wealth have the means to invest in new sources of creating wealth or to otherwise leverage the accumulation of wealth, thus are the beneficiaries of the new wealth. Over time, wealth condensation can significantly contribute to the persistence of inequality within society. Piketty in his book *Capital in the 21st Century* argues that the fundamental force for divergence is the usually greater return of capital (r) than economic growth (g), and that larger fortunes generate higher returns.

Rent Seeking

Economist Joseph Stiglitz argues that rather than explaining concentrations of wealth and income, market forces should serve as a brake on such concentration, which may better be explained by the non-market force known as "rent-seeking". While the market will bid up compensation for rare and desired skills to reward wealth creation, greater productivity, etc., it will also prevent successful entrepreneurs from earning excess profits by fostering competition to cut prices, profits and large compensation. A better explainer of growing inequality, according to Stiglitz, is the use of political power generated by wealth by certain groups to shape government policies financially beneficial to them. This process, known to economists as rent-seeking, brings income not from creation of wealth but from "grabbing a larger share of the wealth that would otherwise have been produced without their effort"

Rent seeking is often thought to be the province of societies with weak institutions and weak rule of law, but Stiglitz believes there is no shortage of it in developed societies such as the United States. Examples of rent seeking leading to inequality include

- the obtaining of public resources by "rent-collectors" at below market prices (such as granting public land to railroads, or selling mineral resources for a nominal price in the US),
- selling services and products to the public at above market prices (medicare drug benefit in the US that prohibits government from negotiating prices of drugs with the drug companies, costing the US government an estimated $50 billion or more per year),
- securing government tolerance of monopoly power (The richest person in the world in 2011, Carlos Slim, controlled Mexico's newly privatized telecommunication industry).

Since rent seeking aims to "pluck the goose to obtain the largest amount of feathers with the least possible amount of hissing" – it is by nature obscure, avoiding public spotlight in legal fine print, or camouflaged its extraction with widely accepted rationalizations (markets are naturally competitive and so need no government regulation against monopolies).

Finance Industry

Jamie Galbraith argues that countries with larger financial sectors have greater inequality, and the link is not an accident.

Mitigating Factors

Countries with a left-leaning legislature have lower levels of inequality. Many factors constrain economic inequality – they may be divided into two classes: government sponsored, and market driven. The relative merits and effectiveness of each approach is a subject of debate.

Typical government initiatives to reduce economic inequality include:

- Public education: increasing the supply of skilled labour and reducing income inequality due to education differentials.
- Progressive taxation: the rich are taxed proportionally more than the poor, reducing the amount of income inequality in society if the change in taxation does not cause changes in income.

Market forces outside of government intervention that can reduce economic inequality include:

- propensity to spend: with rising wealth & income, a person may spend more. In an extreme example, if one person owned everything, they would immediately need to hire people to maintain their properties, thus reducing the wealth concentration.

Effects

Effects of inequality researchers have found include higher rates of health and social problems, and lower rates of social goods, a lower level of economic utility in society from resources devoted on high-end consumption, and even a lower level of economic growth when human capital is neglected for high-end consumption. For the top 21 industrialised countries, counting each person equally, life expectancy is lower in more unequal countries (r = -.907). A similar relationship exists among US states (r = -.620).

2013 Economics Nobel prize winner Robert J. Shiller said that rising inequality in the United States and elsewhere is the most important problem. Increasing inequality harms economic growth. High and persistent unemployment, in which inequality increases, has a negative effect on subsequent long-run economic growth. Unemployment can harm growth not only because it is a waste of resources, but also because it generates redistributive pressures and subsequent distortions, drives people to poverty, constrains liquidity limiting labour mobility, and erodes self-esteem promoting social dislocation, unrest and conflict. Policies aiming at controlling unemployment and in particular at reducing its inequality-associated effects support economic growth.

The economic stratification of society into "elites" and "masses" played a central role in the collapse of other advanced civilizations such as the Roman, Han and Gupta empires.

Health and Social Cohesion

British researchers Richard G. Wilkinson and Kate Pickett have found higher rates of health and social problems (obesity, mental illness, homicides, teenage births, incarceration, child conflict, drug use), and lower rates of social goods (life expectancy by country, educational performance, trust among strangers, women's status, social mobility, even numbers of patents issued) in countries and states with higher inequality. Using statistics from 23 developed countries and the 50 states of the US, they found social/health problems lower in countries like Japan and Finland and states like Utah and New Hampshire with high levels of equality, than in countries (US and UK) and states (Mississippi and New York) with large differences in household income.

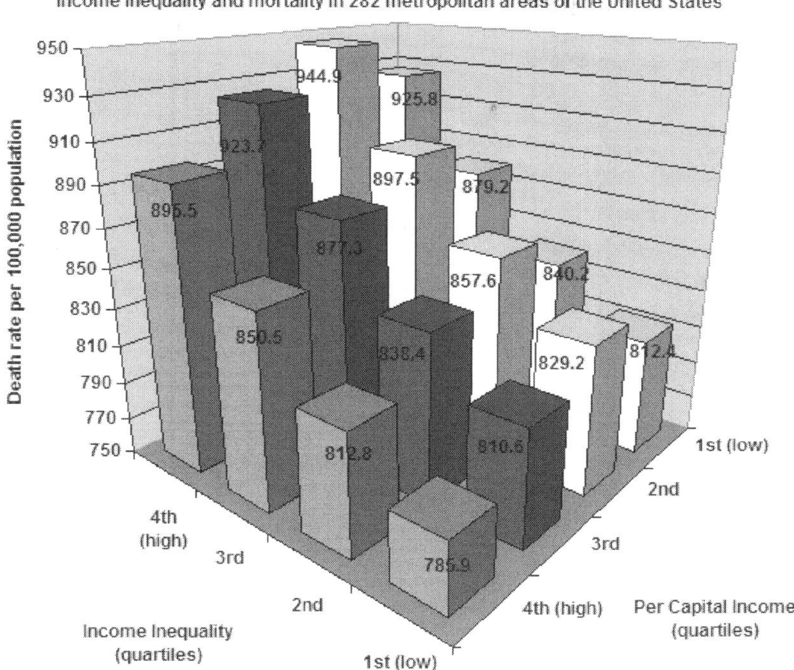

***Figure:** Income inequality and mortality in 282 metropolitan areas of the United States. Mortality is strongly associated with higher income inequality, but, within levels of income inequality, not with per capita income.*

For most of human history higher material living standards – full stomachs, access to clean water and warmth from fuel – led to better health and longer lives. This pattern of higher incomes-longer lives still holds among poorer countries, where life expectancy increases rapidly as per capita income increases, but in recent decades it has slowed down among middle income countries and plateaued among the richest thirty or so countries in the world. Americans live no longer on average (about 77 years in 2004) than Greeks (78 years) or New Zealanders (78), though the USA has a higher GDP per capita. Life expectancy in Sweden (80 years) and Japan (82) – where income was more equally distributed – was longer.

In recent years the characteristic that has strongly correlated with health in developed countries is income inequality. Creating an index of "Health and Social Problems" from nine factors, authors Richard Wilkinson and Kate Pickett found health and social problems "more common in countries with bigger income inequalities", and more common among states in the US with larger income inequalities. Other studies have confirmed this relationship. The UNICEF index of "child well-being in rich countries", studying 40 indicators in 22 countries, correlates with greater equality but not per capita income.

Pickett and Wilkinson argue that inequality and social stratification lead to higher levels of psychosocial stress and status anxiety which can lead to depression, chemical dependency, less community life, parenting problems and stress-related diseases.

Social Cohesion

Research has shown an inverse link between income inequality and social cohesion. In more equal societies, people are much more likely to trust each other, measures of social capital (the benefits of goodwill, fellowship, mutual sympathy and social connectedness among groups who make up a social units) suggest greater community involvement, and homicide rates are consistently lower.

Comparing results from the question "would others take advantage of you if they got the chance?" in U.S General Social Survey and statistics on income inequality, Eric Uslaner and Mitchell Brown found there is a high correlation between the amount of trust in society and the amount of income equality. A 2008 article by Andersen and Fetner also found a strong relationship between economic inequality within and across countries and tolerance for 35 democracies.

In two studies Robert Putnam established links between social capital and economic inequality. His most important studies established these links in both the United States and in Italy. His explanation for this relationship is that

> *Community and equality are mutually reinforcing... Social capital and economic inequality moved in tandem through most of the twentieth century. In terms of the distribution of wealth and income, America in the 1950s and 1960s was more egalitarian than it had been in more than a century... Those same decades were also the high point of social connectedness and civic engagement. Record highs in equality and social capital coincided. Conversely, the last third of the twentieth century was a time of growing inequality and eroding social capital... The timing of the two trends is striking: somewhere around 1965–70 America reversed course and started becoming both less just economically and less well connected socially and politically.*

Albrekt Larsen has advanced this explanation by a comparative study of how trust increased in Denmark and Sweden in the latter part of the 20th century while it decreased in the US and UK. It is argued that inequality levels influence how citizens imagine the trustworthiness of fellow citizens. In this model social trust is not about relations to people you meet (as in Putnam's model) but about people you imagine.

The economist Joseph Stiglitz has argued that economic inequality has led to distrust of business and government.

Crime

Crime rate has also been shown to be correlated with inequality in society. Most studies looking into the relationship have concentrated on homicides – since homicides are almost identically defined across all nations and jurisdictions. There have been over fifty studies showing tendencies for violence to be more common in societies where income differences are larger. Research has been conducted comparing developed countries with undeveloped countries, as well as studying areas within countries. Daly et al. 2001 found that among U.S States and Canadian Provinces there is a tenfold difference in homicide rates related to inequality. They estimated that about half of all variation in homicide rates can be accounted for by differences in the amount of inequality in each province or state. Fajnzylber et al. (2002) found a similar relationship worldwide. Among comments in academic literature on the relationship between homicides and inequality are:

- The most consistent finding in cross-national research on homicides has been that of a positive association between income inequality and homicides.
- Economic inequality is positively and significantly related to rates of homicide despite an extensive list of conceptually relevant controls. The fact that this relationship is found with the most recent data and using a different measure of economic inequality from previous research, suggests that the finding is very robust.

Social, Cultural, and Civic Participation

Higher income inequality led to less of all forms of social, cultural, and civic participation among the less wealthy. When inequality is higher the poor do not shift to less expensive forms of participation.

Utility, Economic Welfare, and Distributive Efficiency

Following the utilitarian principle of seeking the greatest good for the greatest number – economic inequality is problematic. A house that provides less utility to a millionaire as a summer home than it would to a homeless family of five, is an example of reduced "distributive efficiency" within society, that decreases marginal utility of wealth and thus the sum total of personal utility. An additional dollar spent by a poor person will go to things providing a great deal of utility to that person, such as basic necessities like food, water, and healthcare; while, an additional dollar spent by a much richer person will very likely go to luxury items providing relatively less utility to that person. Thus, the marginal utility of wealth per person ("the additional dollar") decreases as a person becomes richer. From this standpoint, for any given amount of wealth in society, a society with more equality will have higher aggregate utility. Some studies have found evidence for this theory, noting that in societies where inequality is lower, population-wide satisfaction and happiness tend to be higher.

Economist Arthur Cecil Pigou Argues that

... it is evident that any transference of income from a relatively rich man to a relatively poor man of similar temperament, since it enables more intense wants, to be satisfied at the expense of less intense wants, must increase the aggregate sum of satisfaction. The old "law of diminishing utility" thus leads securely to the proposition: Any cause which increases the absolute share of real income in the hands of the poor, provided that it does not lead to a contraction in the size of the national dividend from any point of view, will, in general, increase economic welfare.

Philosopher David Schmidtz argues that maximizing the sum of individual utilities will harm incentives to produce.

A society that takes Joe Rich's second unit is taking that unit away from someone who . . . has nothing better to do than plant it and giving it to someone who . . . does have something better to do with it. That sounds good, but in the process, the society takes seed corn out of production and diverts it to food, thereby cannibalizing itself.

However, in addition to the diminishing marginal utility of unequal distribution, Pigou and others point out that a "keeping up with the Joneses" effect among the well off may lead to greater inequality and use of resources for *no* greater return in utility.

A larger proportion of the satisfaction yielded by the incomes of rich people comes from their relative, rather than from their absolute, amount. This part of it will not be destroyed if the incomes of all rich people are diminished together. The loss of economic welfare suffered by the rich when command over resources is transferred from them to the poor will, therefore, be substantially smaller relatively to the gain of economic welfare to the poor than a consideration of the law of diminishing utility taken by itself suggests.

When the goal is to own the biggest yacht – rather than a boat with certain features – there is no greater benefit from owning 100 metre long boat than a 20 m one as long as it is bigger than your rival. Economist Robert H. Frank compare the situation to that of male elks who use their antlers to spar with other males for mating rights.

The pressure to have bigger ones than your rivals leads to an arms race that consumes resources that could have been used more efficiently for other things, such as fighting off disease. As a result, every male ends up with a cumbersome and expensive pair of antlers, ... and "life is more miserable for bull elk as a group."

Aggregate Demand, Consumption and Debt

Income inequality lowers aggregate demand, leading to increasingly large segments of formerly middle class consumers unable to afford as many luxury and essential goods and services. This pushes production and overall employment down.

Conservative researchers have argued that income inequality is not significant because consumption, rather than income should be the measure of inequality, and inequality of consumption is less extreme than inequality of income in the US. Will Wilkinson of the libertarian Cato Institute states that "the weight of the evidence shows that the run-up in consumption inequality has been considerably less dramatic than the rise in income inequality," and consumption is more important than income. According to Johnson, Smeeding, and Tory, consumption inequality was actually lower in 2001 than it was in 1986. The debate is summarized in "The Hidden Prosperity of the Poor" by journalist Thomas B. Edsall. Other studies have not found consumption inequality less dramatic than household income inequality, and the CBO's study found consumption data not "adequately" capturing "consumption by high-income households" as it does their income, though it did agree that household consumption numbers show more equal distribution than household income.

Others dispute the importance of consumption over income, pointing out that if middle and lower income are consuming more than they earn it is because they are saving less or going deeper into debt. Income inequality has been the driving factor in the growing household debt, as high earners bid up the price of real estate and middle income earners go deeper into debt trying to maintain what once was a middle class lifestyle.

Central Banking economist Raghuram Rajan argues that "systematic economic inequalities, within the United States and around the world, have created deep financial 'fault lines' that have made financial crises more likely to happen than in the past" – the Financial crisis of 2007–08 being the most recent example. To compensate for stagnating and declining purchasing power, political pressure has developed to extend easier credit to the lower and middle income earners – particularly to buy homes – and easier credit in general to keep unemployment rates low. This has given the American economy a tendency to go "from bubble to bubble" fuelled by unsustainable monetary stimulation.

Monopolization of Labour, Consolidation, and Competition

Greater income inequality can lead to monopolization of the labour force, resulting in fewer employers requiring fewer workers. Remaining employers can consolidate and take advantage of the relative lack of competition, leading to less consumer choice, market abuses, and relatively higher real prices.

Economic Incentives

Some modern economic theories, such as the neoclassical school, have suggested that a functioning economy entails a certain level of unemployment. These theories argue that unemployment benefits must be below the wage level to provide an incentive to work, thereby mandating inequality. Such theories state additionally that the unemployment rate cannot reduce to zero.

Many economists believe that one of the main reasons that inequality might induce economic incentive is because material well-being and conspicuous consumption relate to status. In this view, high stratification of income (high inequality) creates high amounts of social stratification, leading to greater competition for status.

One of the first writers to note this relationship, Adam Smith, recognised "regard" as one of the major driving forces behind economic activity. From *The Theory of Moral Sentiments* in 1759:

> *What is the end of avarice and ambition, of the pursuit of wealth, of power, and pre-eminence? Is it to supply the necessities of nature? The wages of the meanest laborer can supply them... Why should those who have been educated in the higher ranks of life, regard it as worse than death, to be reduced to live, even without labour, upon the same simple fare with him, to dwell under the same lowly roof, and to be clothed in the same*

> *humble attire? From whence, then, arises that emulation which runs through all the different ranks of men, and what are the advantages which we propose by that great purpose of human life which we call bettering our condition? To be observed, to be attended, to be taken notice of with sympathy, complacency, and approbation, are all the advantages which we can propose to derive from it. It is the vanity, not the ease, or the pleasure, which interests us.*

Modern sociologists and economists such as Juliet Schor and Robert H. Frank have studied the extent to which economic activity is fuelled by the ability of consumption to represent social status. Schor, in *The Overspent American*, argues that the increasing inequality during the 1980s and 1990s strongly accounts for increasing aspirations of income, increased consumption, decreased savings, and increased debt.

In the book *Luxury Fever*, Robert H. Frank argues that satisfaction with levels of income is much more strongly affected by how someone's income compares with others than its absolute level. Frank gives the example of instructions to a yacht architect by a customer – shipping magnate Stavros Niarchos – to make Niarchos' new yacht 50 feet longer than that of rival magnate Aristotle Onassis. Niarchos did not specify or reportedly even know the exact length of Onassis's yacht.

Economic Growth

In the 1960s, economist Arthur Melvin Okun argued that there was a "trade-off" between economic growth and equality. Pursuing equality could reduce efficiency (the total output produced with given resources) by reducing incentives to work, save, and invest and through the "leaky bucket" of wasteful government efforts to redistribute (such as a progressive tax code and minimum wages). Some resources "will simply disappear in transit, so the poor will not receive all the money that is taken from the rich". Along the same lines, earlier writers had argued that wealthier individuals save proportionally more of their incomes, so that more inequality would lead to higher overall savings and thus capital accumulation and growth. On the other side Ozan Hatipoglu argues that is important to reduce the income inequality since "Inequality between rich and poor plays an important role for technological progress, because by determining who are able to afford newer goods, it affects incentives for innovative activity."

Cross-Country Evidence

Many authors have empirically examined the relationship between economic growth and income inequality in a large group of countries.

Following the broader economic growth literature, the typical approach was to relate countries' real GDP per capita growth over a long period of time (e.g., 1965 through 1990) to the income distribution at the start of the period, simultaneously taking into account other standard determinants such as the initial level of real GDP per capita. A typical conclusion was that more unequal countries tend to grow slower (Alesina and Rodrik, 1994), though the evidence was contested.

Because of general dissatisfaction with the empirical approach, including difficulties in determining causality and capturing country-specific factors, attention turned to the analysis of how changes in the income distribution affected the growth rate in subsequent time period (usually five years) in a large group of countries. Forbes (2000) found that an increase in inequality tends to raise growth during the subsequent period. This literature did not go too far as Banerjee and Duflo (2003) found a complex relationship between inequality and growth, in which changes in inequality in either direction lowered growth subsequently. They interpreted this finding as supporting the notion that redistribution hurts growth, at least over the short- to medium-run, but also cautioned about interpreting income distribution-economic growth analysis of this type.

In recent years, the economic growth literature has recognised that growth in most countries does not follow a smooth path, but is characterized by sharp turning points – periods of sustained growth and stagnation. The interesting empirical questions, then, are about the determinants of the turning points (Pritchett, 2000).

Along these lines, Andrew Berg and Jonathan D. Ostry (2011) examined the question of what sustains long periods of strong growth, and found that one of the most robust and important determinants is the level of income inequality. In particular, they found that high 'growth spells' were much more likely to end in countries with less equal income distribution, and that the measured effect was large. For example, they estimate that closing half the inequality gap between Latin America and emerging Asia would more than double the expected duration of a 'growth spell.' Their findings were robust to the inclusion of other variables in the model, and to alternate definitions of growth spells. According to their study, which has featured prominently in the financial press, inequality is of course not the only thing that matters but it clearly belongs in the "pantheon" of well-established growth factors such as the quality of political institutions or trade openness.

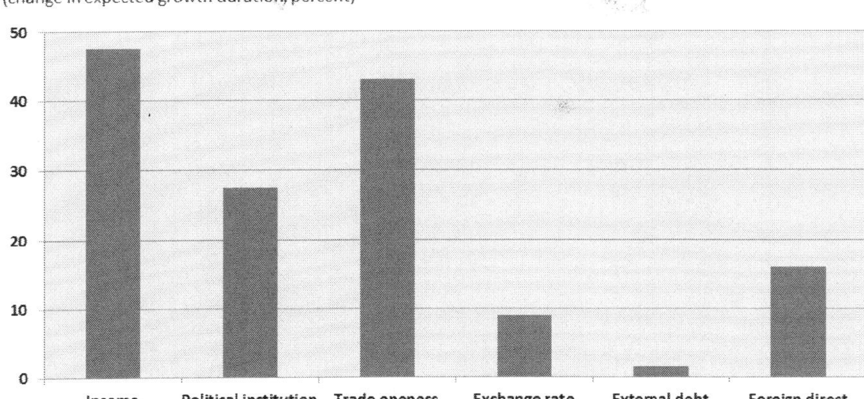

Figure: *Ostry and Berg (2011) studied factors affecting the duration of economic growth in developed and developing countries. They found that income equality has a more beneficial impact than trade openness, sound political institutions, and foreign investment.*

Berg and Ostry postulate that high levels of inequality might damage long term growth by amplifying the potential for financial crisis, discouraging investment because of political instability, making it more difficult for governments to make difficult choices (such as raising taxes or cutting public expenditure) in the face of shocks, or by discouraging investment in education and health for the poor.

Comparisons with the United States

Economic sociologist Lane Kenworthy has found no correlation between levels of inequality and economic growth among developed countries, among states of the US, or in the US over the years from 1947 to 2005. Nor did Jared Bernstein find a correlation, plotting yearly real GDP growth and the share of income going to the top 1%, 1929–2010

Mechanisms

According to economist Branko Milanovic, while traditionally economists thought inequality was good for growth

"The view that income inequality harms growth – or that improved equality can help sustain growth – has become more widely held in recent years. ... The main reason for this shift is the increasing importance of human capital in development. When physical capital mattered most, savings and investments were key. Then it was important to have a large contingent of rich people who could save a greater proportion of their income than the poor and invest it in physical capital. But now that human capital is scarcer than machines, widespread education has become the secret to growth."

"Broadly accessible education" is both difficult to achieve when income distribution is uneven and tends to reduce "income gaps between skilled and unskilled labour."

A study by Perotti (1996) examines of the channels through which inequality may affect economic growth. He shows that in accordance with the credit market imperfection approach, inequality is associated with lower level of human capital formation (education, experience, apprenticeship) and higher level of fertility, while lower level of human capital is associated with lower growth and lower levels of economic growth. In contrast, his examination of the political economy channel refutes the political economy mechanism. He demonstrates that inequality is associated with lower levels of taxation, while lower levels of taxation, contrary to the theories, are associated with lower level of economic growth

The credit market imperfection approach, developed by Galor and Zeira (1993), demonstrates that inequality in the presence of credit market imperfections has a long lasting detrimental effect on human capital formation and economic development.

The political economy approach, developed by Alesian and Rodrik (1994) and Persson and Tabellini (1994), argues that inequality is harmful for economic development because inequality generates a pressure to adopt redistributive policies that have an adverse effect on investment and economic growth.

The sovereign-debt economic problems of the late twenty-oughts do not seem to be correlated to redistribution policies in Europe. With the exception of Ireland, the countries at risk of default in 2011 (Greece, Italy, Spain, Portugal) were notable for their high Gini-measured levels of income inequality compared to other European countries. As measured by the Gini index, Greece as of 2008 had more income inequality than the economically healthy Germany.

Housing

A number of researchers (David Rodda, Jacob Vigdor, and Janna Matlack), argue that a shortage of affordable housing – at least in the US – is caused in part by income inequality. David Rodda noted that from 1984 and 1991, the number of quality rental units decreased as the demand for higher quality housing increased (Rhoda 1994:148). Through gentrification of older neighbourhoods, for example, in East New York, rental prices increased rapidly as landlords found new residents willing to pay higher market rate for housing and left lower income families without rental units. The ad valorem property tax policy combined with rising prices made it difficult or impossible for low income residents to keep pace.

Aspirational Consumption and Household Risk

Firstly, certain costs are difficult to avoid and are shared by everyone, such as the costs of housing, pensions, education and health care. If the state does not provide these services, then for those on lower incomes, the costs must be borrowed and often those on lower incomes are those who are worse equipped to manage their finances. Secondly, aspirational consumption describes the process of middle income earners aspiring to achieve the standards of living enjoyed by their wealthier counterparts and one method of achieving this aspiration is by taking on debt. The result leads to even greater inequality and potential economic instability.

Poverty

Oxfam asserts that worsening inequality is impeding the fight against global poverty. A 2013 report from the group stated that the $240 billion added to the fortunes of the world's richest billionaires in 2012 was enough to end extreme poverty four times over. Oxfam Executive Director Jeremy Hobbs said that "We can no longer pretend that the creation of wealth for a few will inevitably benefit the many – too often the reverse is true."

Jared Bernstein and Elise Gould of the Economic Policy Institute suggest that poverty in the United States could have been significantly mitigated if inequality had not increased over the last few decades.

Perspectives

Socialism and Marxism: Socialists attribute the vast disparities in wealth and income to the private ownership of the means of production by a class of owners, resulting in a situation where a small portion of the population receives unearned income in the form of property income by virtue of ownership titles in capital equipment, financial assets and corporate stock. In contrast, the vast majority of the population is dependent on income in the form of a wage or salary. In order to rectify this situation, socialists argue that the means of production should be publicly owned, so that income differentials would be reflective of individual contribution to the social product.

Marxists ultimately predict the emergence of a communist society based on the common ownership of the means of production, where each individual citizen would have free access to the articles of consumption (*From each according to his ability, to each according to his need*). According to Marxist philosophy, equality in this sense is essential for freedom because equal access to the output of the means of production frees individuals from dependent relationships, allowing them to transcend alienation.

Meritocracy

Meritocracy favours an eventual society where an individual's success is a direct function of his merit, or contribution. Economic inequality would be a natural consequence of the wide range in individual skill, talent and effort in human population.

Liberal Perspectives

Most modern social liberals, including centrist or left-of-centre political groups, believe that the capitalist economic system should be fundamentally preserved, but the status quo regarding the income gap must be reformed. Social liberals favour a capitalist system with active Keynesian macroeconomic policies, neoliberalism, and progressive taxation (to even out differences in income inequality).

However, contemporary classical liberals and libertarians generally do not take a stance on wealth inequality, but believe in equality under the law regardless of whether it leads to unequal wealth distribution. In 1966 Ludwig von Mises, a prominent figure in the Austrian School of economic thought, explains:

The liberal champions of equality under the law were fully aware of the fact that men are born unequal and that it is precisely their inequality that generates social cooperation and civilization. Equality under the law was in their opinion not designed to correct the inexorable facts of the universe and to make natural inequality disappear. It was, on the contrary, the device to secure for the whole of mankind the maximum of benefits it can derive from it. Henceforth no man-made institutions should prevent a man from attaining that station in which he can best serve his fellow citizens.

Robert Nozick argued that government redistributes wealth by force (usually in the form of taxation), and that the ideal moral society would be one where all individuals are free from force. However, Nozick recognised that some modern economic inequalities were the result of forceful taking of property, and a certain amount of redistribution would be justified to compensate for this force but not because of the inequalities themselves. John Rawls argued in *A Theory of Justice* that inequalities in the distribution of wealth are only justified when they improve society as a whole, including the poorest members. Rawls does not discuss the full implications of his theory of justice. Some see Rawls's argument as a justification for capitalism since even the poorest members of society theoretically benefit from increased innovations under capitalism; others believe only a strong welfare state can satisfy Rawls's theory of justice.

Classical liberal Milton Friedman believed that if government action is taken in pursuit of economic equality then political freedom would suffer. In a famous quote, he said:

A society that puts equality before freedom will get neither. A society that puts freedom before equality will get a high degree of both.

Economist Tyler Cowen has argued that though income inequality has increased within nations, globally it has fallen over the last 20 years. He argues that though income inequality may make individual nations worse off, overall, the world has improved as global inequality has been reduced.

Social Justice Arguments

Patrick Diamond and Anthony Giddens (professors of Economics and Sociology, respectively) hold that 'pure meritocracy is incoherent because, without redistribution, one generation's successful individuals would become the next generation's embedded caste, hoarding the wealth they had accumulated'.

They also state that social justice requires redistribution of high incomes and large concentrations of wealth in a way that spreads it more widely, in order to "recognise the contribution made by all sections of the community to building the nation's wealth." (Patrick Diamond and Anthony Giddens, June 27, 2005, New Statesman)

Pope Francis stated in his *Evangelii Gaudium*, that "as long as the problems of the poor are not radically resolved by rejecting the absolute autonomy of markets and financial speculation and by attacking the structural causes of inequality, no solution will be found for the world's problems or, for that matter, to any problems." He later declared that "inequality is the root of social evil."

When income inequality is low, aggregate demand will be relatively high, because more people who want ordinary consumer goods and services will be able to afford them, while the labour force will not be as relatively monopolized by the wealthy.

Effects on Social Welfare

In most western democracies, the desire to eliminate or reduce economic inequality is generally associated with the political left. One practical argument in favour of reduction is the idea that economic inequality reduces social cohesion and increases social unrest, thereby weakening the society.

There is evidence that this is true and it is intuitive, at least for small face-to-face groups of people. Alberto Alesina, Rafael Di Tella, and Robert MacCulloch find that inequality negatively affects happiness in Europe but not in the United States.

It has also been argued that economic inequality invariably translates to political inequality, which further aggravates the problem. Even in cases where an increase in economic inequality makes nobody economically poorer, an increased inequality of resources is disadvantageous, as increased economic inequality can lead to a power shift due to an increased inequality in the ability to participate in democratic processes.

Capabilities Approach

The capabilities approach – sometimes called the human development approach – looks at income inequality and poverty as form of "capability deprivation". Unlike neoliberalism, which "defines well-being as utility maximization", economic growth and income are considered a means to an end rather than the end itself. Its goal is to "widen people's choices and the level of their achieved well-being" through increasing functionings (the things a person values doing), capabilities (the freedom to enjoy functionings) and agency (the ability to pursue valued goals).

When a person's capabilities are lowered, they are in some way deprived of earning as much income as they would otherwise. An old, ill man cannot earn as much as a healthy young man; gender roles and customs may prevent a woman from receiving an education or working outside the home. There may be an epidemic that causes widespread panic, or there could be rampant violence in the area that prevents people from going to work for fear of their lives. As a result, income and economic inequality increases, and it becomes more difficult to reduce the gap without additional aid. To prevent such inequality, this approach believes it's important to have political freedom, economic facilities, social opportunities, transparency guarantees, and protective security to ensure that people aren't denied their functionings, capabilities, and agency and can thus work towards a better relevant income.

Policy Responses Intended to Mitigate

Progressive taxation reduces absolute income inequality when the higher rates on higher-income individuals are paid and not evaded, and transfer payments and social safety nets result in progressive government spending. Wage ratio legislation has also been proposed as a means of reducing income inequality. The OECD asserts that public spending is vital in reducing the ever expanding wealth gap.

The economists Emmanuel Saez and Thomas Piketty recommend much higher top marginal tax rates on the wealthy, up to 50 percent, or 70 percent or even 90 percent. Ralph Nader, Jeffrey Sachs, the United Front Against Austerity, among others, call for a financial transactions tax (also known as the Robin Hood tax) to bolster the social safety net and the public sector.

The Economist wrote in December 2013: "A minimum wage, providing it is not set too high, could thus boost pay with no ill effects on jobs....America's federal minimum wage, at 38% of median income, is one of the rich world's lowest. Some studies find no harm to employment from federal of state minimum wages, others see a small one, but none finds any serious damage."

General limitations on and taxation of rent-seeking are popular across the political spectrum.

Public policy responses addressing causes and effects of income inequality in the US include: progressive tax incidence adjustments, strengthening social safety net provisions such as Aid to Families with Dependent Children, welfare, the food stamp program, Social Security, Medicare, and Medicaid, increasing and reforming higher education subsidies, increasing infrastructure spending, and placing limits on and taxing rent-seeking.

Socioeconomic Status

Socioeconomic status (SES) is an economic and sociological combined total measure of a person's work experience and of an individual's or family's economic and social position in relation to others, based on income, education, and occupation. When analyzing a family's SES, the household income, earners' education, and occupation are examined, as well as combined income, versus with an individual, when their own attributes are assessed.

Socioeconomic status is typically broken into three categories (high SES, middle SES, and low SES) to describe the three areas a family or an individual may fall into. When placing a family or individual into one of these categories, any or all of the three variables (income, education, and occupation) can be assessed.

Additionally, low income and little education have shown to be strong predictors of a range of physical and mental health problems, including respiratory viruses, arthritis, coronary disease, and schizophrenia. These may be due to environmental conditions in their workplace, or, in the case of mental illnesses, may be the entire cause of that person's social predicament to begin with.

Education in higher socioeconomic families is typically stressed as a more important in topic in the household and local community. In poorer areas, where food and safety are priority, education can take a backseat. Youth audiences are particularly at risk for many health and social issues in the United States, such as unwanted pregnancies, drug abuse, and obesity.

Main Factors

Income: *Income* refers to wages, salaries, profits, rents, and any flow of earnings received. Income can also come in the form of unemployment or workers compensation, social security, pensions, interests or dividends, royalties, trusts, alimony, or other governmental, public, or family financial assistance.

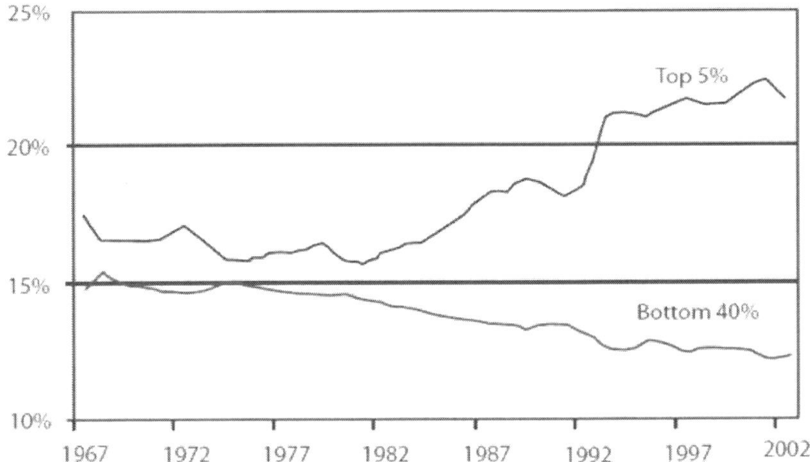

Figure: *This graph shows the differences in income between the top and bottom income earners.*

Income can be looked at in two terms, relative and absolute. Absolute income, as theorized by economist John Maynard Keynes, is the relationship in which as income increases, so will consumption, but not at the same rate. Relative income dictates a person or family's savings and consumption based on the family's income in relation to others. Income is a commonly used measure of SES because it is relatively easy to figure for most individuals.

Income inequality is most commonly measured around the world by the Gini coefficient, where 0 corresponds to perfect equality and 1 means perfect inequality. Low income families focus on meeting immediate needs and do not accumulate wealth that could be passed on to future generations, thus increasing inequality. Families with higher and expendable income can accumulate wealth and focus on meeting immediate needs while being able to consume and enjoy luxuries and weather crises.

Education: Education also plays a role in income. Median earnings increase with each level of education. As conveyed in the chart, the highest degrees, professional and doctoral degrees, make the highest weekly earnings while those without a high school diploma earn less. Higher levels of education are associated with better economic and psychological outcomes (i.e.: more income, more control, and greater social support and networking).

Education plays a major role in skill sets for acquiring jobs, as well as specific qualities that stratify people with higher SES from lower SES. Annette Lareau speaks on the idea of concerted cultivation, where middle class parents take an active role in their children's education and development by using controlled organised activities and fostering a sense of entitlement through encouraged discussion. Laureau argues that families with lower income do not participate in this movement, causing their children to have a sense of constraint. An interesting observation that studies have noted is that parents from lower SES households are more likely to give orders to their children in their interactions while parents with a higher SES are more likely to interact and play with their children. A division in education attainment is thus born out of these two differences in child rearing. Research has shown how children who are born in lower SES households have weaker language skills compared to children raised in higher SES households. These language skills affect their abilities to learn and thus exacerbate the problem of education disparity between low and high SES neighbourhoods. Lower income families can have children who do not succeed to the levels of the middle income children, who can have a greater sense of entitlement, be more argumentative, or be better prepared for adult life.

Research shows that lower SES students have lower and slower academic achievement as compared with students of higher SES. When teachers make judgments about students based on their class and SES, they are taking the first step in preventing students from having an equal opportunity for academic achievement. Educators need to help overcome the stigma of poverty. A student of low SES and low self-esteem should not be reinforced by educators. Teachers need to view students as individuals and not as a member of an SES group. Teachers looking at students in this manner will help them to not be prejudiced towards students of certain SES groups. Raising the level of instruction can help to create equality in student achievement. Teachers relating the content taught to students' prior knowledge and relating it to real world experiences can improve achievement. Educators also need to be open and discuss class and SES differences. It is important that all are educated, understand, and be able to speak openly about SES.

Occupation

Occupational prestige, as one component of SES, encompasses both income and educational attainment. Occupational status reflects the educational attainment required to obtain the job and income levels that vary with different jobs and within ranks of occupations. Additionally, it shows achievement in skills required for the job. Occupational status measures social position by describing job characteristics, decision making ability and control, and psychological demands on the job.

Occupations are ranked by the Census (among other organisations) and opinion polls from the general population are surveyed. Some of the most prestigious occupations are physicians and surgeons, lawyers, chemical and biomedical engineers, university professors, and communications analysts. These jobs, considered to be grouped in the high SES classification, provide more challenging work and greater control over working conditions but require more ability. The jobs with lower rankings include food preparation workers, counter attendants, bartenders and helpers, dishwashers, janitors, maids and housekeepers, vehicle cleaners, and parking lot attendants. The jobs that are less valued also offer significantly lower wages, and often are more laborious, very hazardous, and provide less autonomy.

Occupation is the most difficult factor to measure because so many exist, and there are so many competing scales. Many scales rank occupations based on the level of skill involved, from unskilled to skilled manual labour to professional, or use a combined measure using the education level needed and income involved.

In sum, the majority of researchers agree that income, education and occupation together best represent SES, while some others feel that changes in family structure should also be considered. With the definition of SES more clearly defined, it is now important to discuss the effects of SES on students' cognitive abilities and academic success. Several researchers have found that SES affects students' abilities.

Other Variables

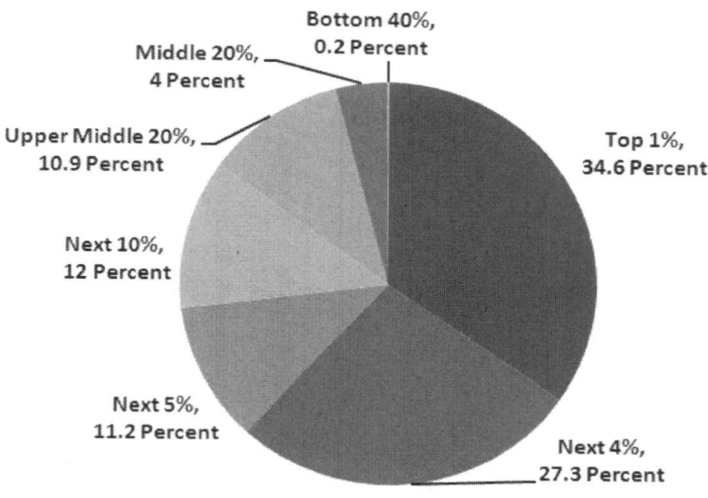

Figure: *This diagram shows the percentages of wealth in the United States in 2007.*

Wealth: Wealth, a set of economic reserves or assets, presents a source of security providing a measure of a household's ability to meet emergencies, absorb economic shocks, or provide the means to live comfortably. Wealth reflects intergenerational transitions as well as accumulation of income and savings.

Income, age, marital status, family size, religion, occupation, and education are all predictors for wealth attainment.

The wealth gap, like income inequality, is very large in the United States. There exists a racial wealth gap due in part to income disparities and differences in achievement resulting from institutional discrimination. According to Thomas Shapiro, differences in savings (due to different rates of incomes), inheritance factors, and discrimination in the housing market lead to the racial wealth gap. Shapiro claims that savings increase with increasing income, but African Americans cannot participate in this, because they make significantly less than whites. Additionally, rates of inheritance dramatically differ between African Americans and whites. The amount a person inherits, either during a lifetime or after death, can create different starting points between two different individuals or families.

These different starting points also factor into housing, education, and employment discrimination. A third reason Shapiro offers for the racial wealth gap are the various discriminations African Americans must face, like redlining and higher interest rates in the housing market. These types of discrimination feed into the other reasons why African Americans end up having different starting points and therefore fewer assets.

Effects

Health: Recently, there has been increasing interest from epidemiologists on the subject of economic inequality and its relation to the health of populations. Socioeconomic status is an important source of health inequity, as there is a very robust positive correlation between socioeconomic status and health. This correlation suggests that it is not only the poor who tend to be sick when everyone else is healthy, but that there is a continual gradient, from the top to the bottom of the socio-economic ladder, relating status to health. This phenomenon is often called the "SES Gradient". Lower socioeconomic status has been linked to chronic stress, heart disease, ulcers, type 2 diabetes, rheumatoid arthritis, certain types of cancer, and premature aging.

There is debate regarding the cause of the SES Gradient. A number of researchers see a definite link between economic status and mortality due to the greater economic resources of the wealthy, but they find little correlation due to social status differences.

Other researchers such as Richard G. Wilkinson, J. Lynch, and G.A. Kaplan have found that socioeconomic status strongly affects health even when controlling for economic resources and access to health care. Most famous for linking social status with health are the Whitehall studies—a series of studies conducted on civil servants in London. The studies found that although all civil servants in England have the same access to health care, there was a strong correlation between social status and health. The studies found that this relationship remained strong even when controlling for health-affecting habits such as exercise, smoking and drinking. Furthermore, it has been noted that no amount of medical attention will help decrease the likelihood of someone getting type 2 diabetes or rheumatoid arthritis—yet both more common among populations with lower socioeconomic status. There is no significant relationship between SES and stress during pregnancy, while there is a significant relationship between husband's occupational status Also, there is not significant relationship between income and mother's education and the rate of pregnancy stress

Political Participation

Political scientists have established a consistent relationship between SES and political participation.

Psychological

Language Development

Home Environment: The environment of low SES children is characterized by less dialogue from parents, minimal amounts of book reading, and few instances of joint attention, the shared focus of the child and adult on the same object or event, when compared to the environment of high SES children. In contrast, infants from high SES families experience more child-directed speech. At 10 months, children of high SES hear on average 400 more words than their low SES peers.

Parental Interactions

In addition to the amount of language input from parents, SES heavily influences the type of parenting style a family chooses to practice. These different parenting styles shape the tone and purpose of verbal interactions between parent and child. For example, parents of high SES tend toward more authoritative or permissive parenting styles. These parents pose more open-ended questions to their children to encourage the latter's speech growth. In contrast, parents of low SES tend toward more authoritarian styles of address. Their conversations with their children contain more imperatives and yes/no questions that inhibits child responses and speech development.

Parental differences in addressing children may be traced to the position of their respective groups within society. Working class individuals often hold low power, subordinate positions in the occupational world. This standing in the social hierarchy requires a personality and interaction style that is relational and capable of adjusting to circumstances. An authoritarian style of address prepares children for these types of roles, which require a more accommodating and compliant personality. Therefore, low SES parents see the family as more hierarchical, with the parents at the top of the power structure, which shapes verbal interaction. This power differential emulates the circumstances of the working class world, where individuals are ranked and discouraged from questioning authority.

Conversely, high SES individuals occupy high power positions that call for greater expressivity. High SES parents encourage their children to question the world around them. In addition to asking their children more questions, these parents push their children to create questions of their own. In contrast with low SES parents, these individuals often view the power disparity between parent and child as detrimental to the family. Opting instead to treat children as equals, high SES conversations are characterized by a give and take between parent and child. These interactions help prepare these children for occupations that require greater expressivity.

Disparities in Language Acquisition

The linguistic environment of low and high SES children differs substantially, which affects many aspects of language and literacy development such as semantics, syntax, morphology, and phonology.

Semantics

Semantics is the study of the meaning of words and phrases. Semantics covers vocabulary, which is affected by SES. Children of high SES have larger expressive vocabularies by the age of 24 months due to more efficient processing of familiar words. By age 3, there are significant differences in the amount of dialogue and vocabulary growth between children of low and high SES. A lack of joint attention in children contributes to poor vocabulary growth when compared to their high SES peers. Joint attention and book reading are important factors that affect children's vocabulary growth. With joint attention, a child and adult can focus on the same object, allowing the child to map out words. For example, a child sees an animal running outside and the mom points to it and says, "Look, a dog." The child will focus its attention to where its mother is pointing and map the word dog to the pointed animal. Joint attention thus facilitates word learning for children.

Syntax

Syntax refers to the arrangement of words and phrases to form sentences. SES affects the production of sentence structures. Although 22- to 44-month-old children's production of simple sentence structures does not vary by SES, low SES does contribute to difficulty with complex sentence structures. Complex sentences include sentences that have more than one verb phrase. An example of a complex sentence is, "I want you to sit there". The emergence of simple sentence structures is seen as a structure that is obligatory in everyday speech. Complex sentence structures are optional and can only be mastered if the environment fosters its development.

This lag in the sentence formation abilities of low SES children may be caused by less frequent exposure to complex syntax through parental speech. Low SES parents ask fewer response-coaxing questions of their children which limits the opportunities of these children to practice more complex speech patterns. Instead, these parents give their children more direct orders, which has been found to negatively influence the acquisition of more difficult noun and verb phrases. In contrast, high SES households ask their children broad questions to cultivate speech development. Exposure to more questions positively contributes to children's vocabulary growth and complex noun phrase constructions.

Morphology

Children's grasp of morphology, the study of how words are formed, is affected by SES. Children of high SES have advantages in applying grammatical rules, such as the pluralization of nouns and adjectives compared to children of low SES. Pluralizing nouns consists of understanding that some nouns are regular and -s denotes more than one, but also understanding how to apply different rules to irregular nouns. Learning and understanding how to use plural rules is an important tool in conversation and writing. In order to communicate successfully that there is more than one dog running down the street, an -s must be added to dog. Research also finds that the gap in ability to pluralize nouns and adjectives does not diminish by age or schooling because low SES children's reaction times to pluralize nouns and adjectives do not decrease.

Phonology

Phonological awareness, the ability to recognise that words are made up different sound units, is also affected by SES. Children of low SES between the second and sixth grades are found to have low phonological awareness. The gap in phonological awareness increases by grade level. This gap is even more problematic if children of low SES are already born with low levels of phonological awareness and their environment does not foster its growth. Children who are have high phonological awareness from an early age are not affected by SES.

Positive Outcomes of Low SES

Given the large amount of research on the set backs children of low SES face, there is a push by child developmental researchers to steer research to a more positive direction regarding low SES. The goal is to highlight the strengths and assets low income families possess in raising children. For example, African American preschoolers of low SES exhibit strengths in oral narrative, or storytelling, that may promote later success in reading. These children have better narrative comprehension when compared to peers of higher SES.

Literacy Development

A gap in reading growth exists between low SES and high SES children, which widens as children move on to higher grades. Reading assessments that test reading growth include measures on basic reading skills (i.e., print familiarity, letter recognition, beginning and ending sounds, rhyming sounds, word recognition), vocabulary (receptive vocabulary), and reading comprehension skills (i.e., listening comprehension, words in context). The reading growth gap is apparent between the spring of kindergarten and the spring of first-grade, the time when children rely more on the school for reading growth and less on their parents. Initially, high SES children begin as better readers than their low SES counterparts. As children get older, high SES children progress more rapidly in reading growth rates than low SES children. These early reading outcomes affect later academic success. The further children fall behind, the more difficult it is to catch up and the more likely they will continue to fall behind. By the time students enter high school in the United States, low SES children are considerably behind their high SES peers in reading growth.

Home Environment

The disparities in experiences in the home environment children of high and low SES affect reading outcomes. The home environment is considered the main contributor to SES reading outcomes. Children of low SES status are read to less often and have fewer books in the home than their high SES peers, which suggests an answer to why children of low SES status have lower initial reading scores than their high SES counterparts upon entering kindergarten.

The home environment makes the largest contribution to the prediction of initial kindergarten reading disparities. Characteristics of the home environment include home literacy environment and parental involvement in school. Home literacy environment is characterized by the frequency with which parents engage in joint book reading with the child, the frequency with which children read books outside of school, and the frequency with which household members visited the library with the child. Parental involvement in school is characterized by attending a parent–teacher conference, attending a parent–teacher association (PTA) meeting, attending an open house, volunteering, participating in fundraising, and attending a school event. Resources, experiences, and relationships associated with the family are most closely associated with reading gaps when students reading levels are first assessed in kindergarten. The influence of family factors on initial reading level may be due to children experiencing little schooling before kindergarten—they mainly have their families to rely on their reading growth.

Family SES is also associated with reading achievement growth during the summer. Students from high SES families continue to grow in their ability to read after kindergarten and students from low SES families fall behind in their reading growth at a comparable amount. Additionally, the summer setback disproportionately affects African American and Hispanic students because they are more likely than White students to come from low SES families. Also, low SES families typically lack the appropriate resources to continue reading growth when school is not in session.

Neighbourhood Influence

The neighbourhood setting in which children grow up in contributes to reading disparities between low and high SES children. These neighbourhood qualities include but are not limited to garbage or litter in the street, individuals selling or using drugs in the street, burglary or robbery in the area, violent crime in the area, vacant homes in the area, and how safe it is to play in the neighbourhood. Low SES children are more likely to grow up in such neighbourhood conditions than their high SES peers. Community support for the school and poor physical conditions surrounding the school are also associated with children's reading. Neighbourhood factors help explain the variation in reading scores in school entry, and especially as children move on to higher grades. As low SES children in poor neighbourhood environments get older, they fall further behind their high SES peers in reading growth and thus have a more difficult time developing reading skills at grade level.

School Influence

School characteristics, including characteristics of peers and teachers, contribute to reading disparities between low and high SES children. For instance, peers play a role in influencing early reading proficiency. In low SES schools, there are higher concentrations of less skilled, lower SES, and minority peers who have lower gains in reading. The number of children reading below grade and the presence of low-income peers were consistently associated with initial achievement and growth rates. Low SES peers tend to have limited skills and fewer economic resources than high SES children, which makes it difficult for children to grow in their reading ability. The most rapid growth of reading ability happens between the spring of kindergarten and the spring of first grade. Teacher experience (number of years teaching at a particular school and the number of years teaching a particular grade level), teacher preparation to teach (based on the number of courses taken on early education, elementary education, and child development), the highest degree earned, and the number of courses taken on teaching reading all determine whether or not a reading teacher is qualified. Low SES students are more likely to have less qualified teachers, which is associated with their reading growth rates being significantly lower than the growth rates of their high SES counterparts.

Influences on Nonverbal Behaviour

Michael Kraus and Dacher Keltner, in their study published in the December 2008 issue of *Psychological Science,* found that children of parents with a high SES tended to express more disengagement behaviours than their peers of low SES. In this context, disengagement behaviours included self-grooming, fidgeting with nearby objects, and doodling while being addressed. In contrast, engagement behaviours included head nods, eyebrow raises, laughter and gazes at one's partner. These cues indicated an interest in one's partner and the desire to deepen and enhance the relationship. Participants of low SES tended to express more engagement behaviours toward their conversational partners, while their high SES counterparts displayed more disengagement behaviours. Authors hypothesized that, as SES rises, the capacity to fulfill one's needs also increases. This may lead to greater feelings of independence, making individuals of high SES less inclined to gain rapport with conversational partners because they are less likely to need their assistance in the future.

Chapter 4

Racism

Racism consists of both prejudice and discrimination based in social perceptions of biological differences between peoples. It often takes the form of social actions, practices or beliefs, or political systems that consider different races to be ranked as inherently superior or inferior to each other, based on presumed shared inheritable traits, abilities, or qualities. It may also hold that members of different races should be treated differently.

Among the questions about how to define racism are the question of whether to include forms of discrimination that are unintentional, such as making assumptions about preferences or abilities of others based on racial stereotypes, whether to include symbolic or institutionalized forms of discrimination such as the circulation of ethnic stereotypes through the media, and whether to include the socio-political dynamics of social stratification that sometimes have a racial component.

In sociology and psychology, some definitions only include consciously malignant forms of discrimination. However, some consider any assumption that a person's behaviour is tied to their racial categorization to be inherently racist, regardless of whether the action is intentionally harmful or pejorative, because stereotyping necessarily subordinates individual identity to group identity. Some definitions of racism also include discriminatory behaviours and beliefs based on cultural, national, ethnic, caste, or religious stereotypes. One view holds that racism is best understood as 'prejudice plus power' because without the support of political or economic power, prejudice would not be able to manifest as a pervasive cultural, institutional or social phenomenon.

While race and ethnicity are considered to be separate phenomena in contemporary social science, the two terms have a long history of equivalence in popular usage and older social science literature. Racism and racial discrimination are often used to describe discrimination on an ethnic or cultural basis, independent of whether these differences are described as racial. According to the United Nations convention, there is no distinction between the terms *racial discrimination* and *ethnic discrimination,* and superiority based on racial differentiation is scientifically false, morally condemnable, socially unjust and dangerous, and that there is no justification for racial discrimination, in theory or in practice, anywhere.

In history, racism was a driving force behind the transatlantic slave trade, and behind states based on racial segregation such as the U.S. in the nineteenth and early twentieth centuries and South Africa under apartheid. Practices and ideologies of racism are universally condemned by the United Nations in the Declaration of Human Rights. It has also been a major part of the political and ideological underpinning of genocides such as The Holocaust, but also in colonial contexts such as the rubber booms in South America and the Congo, and in the European conquest of the Americas and colonization of Africa, Asia and Australia.

Definitions

Racism involves the belief in racial differences, which acts as a justification for non-equal treatment (which some regard as "discrimination") of members of that race. The term is commonly used negatively and is usually associated with race-based prejudice, violence, dislike, discrimination, or oppression, the term can also have varying and contested definitions. *Racialism* is a related term, sometimes intended to avoid these negative meanings.

As a word, racism is an "-ism", a belief that can be described by a word ending in the suffix -ism, pertaining to race. As its etymology would suggest, its usage is relatively recent and as such its definition is not entirely settled. The *Oxford English Dictionary* defines racism as the "belief that all members of each race possess characteristics, abilities, or qualities specific to that race, especially so as to distinguish it as inferior or superior to another race or races" and the expression of such prejudice, while the *Merriam-Webster's Dictionary* defines it as a belief that race is the primary determinant of human traits and capacities and that racial differences produce an inherent superiority or inferiority of a particular racial group, and alternatively that it is also the prejudice based on such a belief. The *Macquarie Dictionary* defines racism as: "the belief that human races have distinctive characteristics which determine their respective cultures, usually involving the idea that one's own race is superior and has the right to rule or dominate others."

Legal

The UN does not define "racism"; however, it does define "racial discrimination": According to the United Nations Convention on the Elimination of All Forms of Racial Discrimination,

> *the term "racial discrimination" shall mean any distinction, exclusion, restriction, or preference based on race, colour, descent, or national or ethnic origin that has the purpose or effect of nullifying or impairing the recognition, enjoyment or exercise, on an equal footing, of human rights and fundamental freedoms in the political, economic, social, cultural or any other field of public life.*

This definition does not make any difference between discrimination based on ethnicity and race, in part because the distinction between the two remains debatable among anthropologists. Similarly, in British law the phrase *racial group* means "any group of people who are defined by reference to their race, colour, nationality (including citizenship) or ethnic or national origin".

In Norway, the word "race" has been removed from national laws concerning discrimination as the use of the phrase is considered problematic and unethical. The Norwegian Anti-Discrimination Act bans discrimination based on ethnicity, national origin, descent and skin colour.

Sociological

Some sociologists have defined racism as a system of categorical privilege. In *Portraits of White Racism*, David Wellman has defined racism as "culturally sanctioned beliefs, which, regardless of intentions involved, defend the advantages whites have because of the subordinated position of racial minorities". Sociologists Noël A. Cazenave and Darlene Alvarez Maddern define racism as "... a highly organised system of 'race'-based group privilege that operates at every level of society and is held together by a sophisticated ideology of colour/ 'race' supremacy. Sellers and Shelton (2003) found that a relationship between racial discrimination and emotional distress was moderated by racial ideology and public regard beliefs. That is, racial centrality appears to promote the degree of discrimination African American young adults perceive whereas racial ideology may buffer the detrimental emotional effects of that discrimination. Racist systems include, but cannot be reduced to, racial bigotry,".

Some sociologists have also argued, with reference to the USA and elsewhere, that forms of racism have in many instances mutated from more blatant expressions hereof into more covert kinds (albeit that blatant forms of hatred and discrimination still endure). The "newer" (more hidden and less easily detectable) forms of racism—which can be considered as embedded in social processes and structures—are more difficult to explore as well as challenge. It has been suggested that, while in many countries overt and explicit racism has become increasingly taboo, even in those who display egalitarian explicit attitudes, an implicit or aversive racism is still maintained subconsciously.

Xenophobia

Dictionary definitions of *xenophobia* include: intense or irrational dislike or fear of people from other countries (Oxford Dictionaries), unreasonable fear and hatred of strangers or foreigners or of anything that is strange or foreign (Merriam-Webster) The *Dictionary of Psychology* defines it as "a fear of strangers".

Supremacism

Centuries of European colonialism of the Americas, Africa and Asia was often justified by white supremacist attitudes. During the early 20th century, the phrase "The White Man's Burden" was widely used to justify imperialist policy as a noble enterprise.

Figure: *A rally against school integration in 1959.*

Segregationism

Racial segregation is the separation of humans into racial groups in daily life. It may apply to activities such as eating in a restaurant, drinking from a water fountain, using a bath room, attending school, going to the movies, or in the rental or purchase of a home. Segregation is generally outlawed, but may exist through social norms, even when there is no strong individual preference for it, as suggested by Thomas Schelling's models of segregation and subsequent work.

Types

Racial Discrimination

Racial discrimination refers to the separation of people through a process of social division into categories not necessarily related to races for purposes of differential treatment. Racial segregation policies may formalize it, but it is also often exerted without being legalized. Researchers Marianne Bertrand and Sendhil Mullainathan, at the University of Chicago and MIT found in a 2004 study that there was widespread discrimination in the workplace against job applicants whose names were merely perceived as "sounding black". These applicants were 50% less likely than candidates perceived as having "white-sounding names" to receive callbacks for interviews. Devah Pager, a sociologist at Princeton University, sent matched pairs of applicants to apply for jobs in Milwaukee and New York City, finding that black applicants received callbacks or job offers at half the rate of equally qualified whites. In contrast, institutions and courts have upheld discrimination against whites when it is done to promote a diverse work or educational environment, even when it was shown to be to the detriment of qualified applicants. The researchers view these results as strong evidence of unconscious biases rooted in the United States' long history of discrimination (e.g., Jim Crow laws, etc.)

Institutional

Institutional racism (also known as structural racism, state racism or systemic racism) is racial discrimination by governments, corporations, religions, or educational institutions or other large organisations with the power to influence the lives of many individuals. Stokely Carmichael is credited for coining the phrase *institutional racism* in the late 1960s. He defined the term as "the collective failure of an organisation to provide an appropriate and professional service to people because of their colour, culture or ethnic origin".

Maulana Karenga argued that racism constituted the destruction of culture, language, religion and human possibility, and that the effects of racism were "the morally monstrous destruction of human possibility involved redefining African humanity to the world, poisoning past, present and future relations with others who only know us through this stereotyping and thus damaging the truly human relations among peoples."

Economic

Historical economic or social disparity is alleged to be a form of discrimination caused by past racism and historical reasons, affecting the present generation through deficits in the formal education and kinds of preparation in previous generations, and through primarily unconscious racist attitudes and actions on members of the general population.

In 2011, Bank of America agreed to pay $335 million to settle a federal government claim that its mortgage division, Countrywide Financial, discriminated against black and Hispanic homebuyers.

During the Spanish colonial period, Spaniards developed a complex caste system based on race, which was used for social control and which also determined a person's importance in society. While many Latin American countries have long since rendered the system officially illegal through legislation, usually at the time of their independence, prejudice based on degrees of perceived racial distance from European ancestry combined with one's socioeconomic status remain, an echo of the colonial caste system.

Symbolic/Modern

Some scholars argue that in the US earlier violent and aggressive forms of racism have evolved into a more subtle form of prejudice in the late 20th century. This new form of racism is sometimes referred to as "modern racism" and characterized by outwardly acting unprejudiced while inwardly maintaining prejudiced attitudes, and displaying subtle prejudist behaviours such as actions informed by attributing qualities to others based on racial stereotypes, and evaluating the same behaviour differently based on the race of the person being evaluated. This view is based on studies of prejudice and discriminatory behaviour, where some people will act ambivalently towards black people, with positive reactions in certain, more public contexts, but more negative views and expressions in more private contexts. This ambivalence may also be visible for example in hiring decisions where job candidates that are otherwise positively evaluated may be unconsciously disfavoured by employers in the final decision because of their race. Some scholars consider modern racism to be characterized by an explicit rejection of stereotypes, combined with resistance to changing structures of discrimination for reasons that are ostensibly non-racial, an idiology that considers opportunity at a purely individual basis denying the relevance of race in determining individual opportunities, and the exhibition of indirect forms of micro-aggression and/or avoidance towards people of other races.

Cultural Racism

Cultural racism is a term used to describe and explain new racial ideologies and practices that have emerged since World War II. It can be defined as societal beliefs and customs that promote the assumption that the products of a given culture, including the language and traditions of that culture are superior to those of other cultures. It shares a great deal with xenophobia, which is often characterised by fear of, or aggression toward, members of an outgroup by members of an ingroup.

Cultural racism exists when there is a widespread acceptance of stereotypes concerning different ethnic or population groups. Where racism can be characterised by the belief that one race is inherently superior to another, cultural racism can be characterised by the belief that one culture is inherently superior to another.

Declarations and International Law against Racial Discrimination

In 1919, a proposal to include a racial equality provision in the Covenant of the League of Nations was supported by a majority, but not adopted in the Paris Peace Conference, 1919. In 1943, Japan and its allies declared work for the abolition of racial discrimination to be their aim at the Greater East Asia Conference. Article 1 of the 1945 UN Charter includes "promoting and encouraging respect for human rights and for fundamental freedoms for all without distinction as to race" as UN purpose.

In 1950, UNESCO suggested in *The Race Question* —a statement signed by 21 scholars such as Ashley Montagu, Claude Lévi-Strauss, Gunnar Myrdal, Julian Huxley, etc. — to "drop the term *race* altogether and instead speak of ethnic groups". The statement condemned scientific racism theories that had played a role in the Holocaust. It aimed both at debunking scientific racist theories, by popularizing modern knowledge concerning "the race question," and morally condemned racism as contrary to the philosophy of the Enlightenment and its assumption of equal rights for all. Along with Myrdal's *An*

American Dilemma: The Negro Problem and Modern Democracy (1944), *The Race Question* influenced the 1954 U.S. Supreme Court desegregation decision in "Brown v. Board of Education of Topeka". Also in 1950, the European Convention on Human Rights was adopted, widely used on racial discrimination issues.

The United Nations use the definition of racial discrimination laid out in the *International Convention on the Elimination of All Forms of Racial Discrimination*, adopted in 1966:

... *any distinction, exclusion, restriction or preference based on race, colour, descent, or national or ethnic origin that has the purpose or effect of nullifying or impairing the recognition, enjoyment or exercise, on an equal footing, of human rights and fundamental freedoms in the political, economic, social, cultural or any other field of public life.* (Part 1 of Article 1 of the U.N. International Convention on the Elimination of All Forms of Racial Discrimination)

In 2001, the European Union explicitly banned racism, along with many other forms of social discrimination, in the Charter of Fundamental Rights of the European Union, the legal effect of which, if any, would necessarily be limited to Institutions of the European Union: "Article 21 of the charter prohibits discrimination on any ground such as race, colour, ethnic or social origin, genetic features, language, religion or belief, political or any other opinion, membership of a national minority, property, disability, age or sexual orientation and also discrimination on the grounds of nationality."

Ideology

As an ideology, racism existed during the 19th century as "scientific racism", which attempted to provide a racial classification of humanity. Johann Blumenbach in 1775, advocating polygenism, divided the world's population into five groups according to skin colour (Caucasians, Mongols, etc.). The archetypical form of racism is, perhaps, found with the polygenist Christoph Meiners. He split mankind into two divisions which he labelled the "beautiful White race" and the "ugly Black race". In Meiners book *The Outline of History of Mankind* he claimed that a main characteristic of race is either beauty or ugliness. He viewed only the white race as beautiful. He considered ugly races as inferior, immoral and animal like.

Anders Retzius next disproved that Blumenbach's polygenism had any fundamental merit, in demonstrating that neither Europeans nor different nations are one "pure race", but of mixed origins. While discredited, derivations of Blumenbach's taxonomy are still widely used for classification of the population in USA. H. P. Steensby, while strongly emphasising that all humans today are of mixed origins, in 1907 claimed that the origins of human differences must be traced extraordinarily far back in time, and conjectured that the "purest race" today would be the Australian Aboriginals.

After the rejection of polygenism as well as the widespread racist and nationalist violence of the 1930s and 1940s, scientific racism fell strongly out of favour, but the origins of fundamental human and societal differences are still researched within academia, in fields such as human genetics including paleogenetics, social anthropology, comparative politics, history of religions, history of ideas, prehistory, history, ethics, and psychiatry. All reject Meiners' blunt racism. Outside USA, there is widespread rejection of any methodology based on anything similar to Blumenbach's races. It is more unclear to which extent ethnic and national stereotypes are accepted, and when.

Although after World War II and the Holocaust, racist ideologies have been widely discredited on ethical, political and scientific grounds, racism and racial discrimination have remained widespread around the world. Some examples of this in present day are statistics including, but not limited to, the racial breakdown of the prison population versus the national population, physical abilities and mental ability statistics, and other data gathered by scientific groups. While these statistics may be accurate, and can show trends, it's inappropriate in most countries to assume that because a particular race has a high crime or low literacy rate, that the entire race of people are inherent criminals, or inherently unintelligent.

It was already noted by Du Bois that, in making the difference between races, it is not race that we think about, but culture: "... a common history, common laws and religion, similar habits of thought and a conscious striving together for certain ideals of life". Late 19th century nationalists were the first to embrace contemporary discourses on "race", ethnicity and "survival of the fittest" to shape new nationalist doctrines. Ultimately, race came to represent not only the most important traits of the human body, but was also regarded as decisively shaping the character and personality of the nation. According to this view, culture is the physical manifestation created by ethnic groupings, as such fully determined by racial characteristics. Culture and race became considered intertwined and dependent upon each other, sometimes even to the extent of including nationality or language to the set of definition. Pureness of race tended to be related to rather superficial characteristics that were easily addressed and advertised, such as blondness. Racial qualities tended to be related to nationality and language rather than the actual geographic distribution of racial characteristics. In the case of Nordicism, the denomination "Germanic" became virtually equivalent to superiority of race.

Bolstered by some nationalist and ethnocentric values and achievements of choice, this concept of racial superiority evolved to distinguish from other cultures, that were considered inferior or impure. This emphasis on culture corresponds to the modern mainstream definition of racism: "Racism does not originate from the existence of 'races'. It *creates* them through a process of social division into categories: anybody can be racialised, independently of their somatic, cultural, religious differences."

This definition explicitly ignores the biological concept of race, still subject to scientific debate. In the words of David C. Rowe "A racial concept, although sometimes in the guise of another name, will remain in use in biology and in other fields because scientists, as well as lay persons, are fascinated by human diversity, some of which is captured by race."

Until recently, this racist abuse of physical anthropology has been politically exploited. Apart from being unscientific, racial prejudice became subject to international legislation. For instance, the Declaration on the Elimination of All Forms of Racial Discrimination, adopted by the United Nations General Assembly on November 20, 1963, address racial prejudice explicitly next to discrimination for reasons of race, colour or ethnic origin (Article I).

Racism has been a motivating factor in social discrimination, racial segregation, hate speech and violence (such as pogroms, genocides and ethnic cleansings). Despite the persistence of racial stereotypes, humor and epithets in much everyday language, racial discrimination is illegal in many countries.

Ironically, anti-racism has also become a political instrument of abuse. In a reversal of values, anti-racism is being propagated by despots in the service of obscurantism and the suppression of women. Philosopher Pascal Bruckner claimed that *"Anti-racism in the UN has become the ideology of totalitarian regimes who use it in their own interests."*

Inter-Minority Variants

Prejudiced thinking among and between minority groups does occur.

In Europe

In Britain, tensions between minority groups can be just as strong as those between minorities and the majority population. In Birmingham, there have been long-term divisions between the Black and South Asian communities, which were illustrated in the Handsworth riots and in the smaller 2005 Birmingham riots. In Dewsbury, a Yorkshire town with a relatively high Muslim population, there have been tensions and minor civil disturbances between Kurds and South Asians.

In France, home to Europe's largest population of Muslims (about 6 million) as well as the continent's largest community of Jews (about 600,000), anti-Jewish violence, property destruction, and racist language has been increasing over the last several years. Jewish leaders perceive the Muslim population as intensifying antisemitism in France, mainly among Muslims of Arab or African heritage, but also this antisemitism is perceived as also growing among Caribbean islanders from former colonies.

In North America

For example, conflicts between African Americans and Korean Americans (notably in the Los Angeles riots of 1992), by blacks towards Jews (such as the riots in Crown Heights in 1991), between new immigrant groups (such as Latinos), or towards whites.

There has been a long-running racial tension between African Americans and Mexican Americans. There have been several significant riots in California prisons in which Mexican American inmates and African Americans have specifically targeted each other based on racial reasons. There have been reports of racially motivated attacks against African Americans who have moved into neighbourhoods occupied mostly by Mexican Americans, and vice versa.

In the late 1920s in California, there was animosity between the Filipinos and the Mexicans and between European Americans and Filipino Americans since they competed for the same jobs. Recently, there has also been an increase in racial violence between African immigrants and Blacks who have already lived in the country for generations.

Over 50 members of the Azusa 13 gang, associated with the Mexican Mafia, were indicted in 2011 for harassing and intimidating African Americans.

Anti-Racism

Anti-racism includes beliefs, actions, movements, and policies adopted or developed to oppose racism. In general, it promotes an egalitarian society in which people are not discriminated against in race. Movements such as the African-American Civil Rights Movement and the Anti-Apartheid Movement were examples of anti-racist movements. Nonviolent resistance is sometimes an element of anti-racial movements, although this was not always the case. Hate crime laws, affirmative action, and bans on racist speech are also examples of government policy designed to suppress racism.

International Day for the Elimination of Racial Discrimination

UNESCO marks March 21 as the yearly International Day for the Elimination of Racial Discrimination, in memory of the events that occurred on March 21, 1960 in Sharpeville, South Africa, where police killed student demonstrators peacefully protesting against the apartheid regime.

Well-Being

Well-being or welfare is a general term for the condition of an individual or group, for example their social, economic, psychological, spiritual or medical state; high well-being means that, in some sense, the individual or group's experience is positive, while low well-being is associated with negative happenings.

In economics, the term is used for one or more quantitative measures intended to assess the quality of life of a group, for example, in the capabilities approach and the economics of happiness. Like the related cognate terms 'wealth' and 'welfare', economics sources may contrast the state with its opposite. The study of well-being is divided into subjective well-being and objective well-being.

Background in Well-Being

Although there has not been a clear definition established for well-being, it can be defined as "a special case for attitude". This definition serves two purposes of well-being: developing and testing a systematic theory for the structure of [interrelationships] among varieties of well-being and integration of well-being theory with the ongoing cumulative theory development in the fields of attitude of related research". One's well-being develops through assessments of their environment and emotions and then developing an interpretation of their own personal self. There are two different types of well-being: cognitive and affective.

Cognitive Well-Being

Cognitive well-being is developed through assessing one's interactions with their environment and other people. "Welfare economics ultimately deals with cognitive concepts such as well-being, happiness, and satisfaction. These relate to notions such as aspirations and needs, contentment and disappointment". People tend to assess their cognitive well-being based on the social classes that are in their community. In communities with a wide variety of social statuses, the lower class will tend to compare their lifestyle to those of higher class and assess what they do and do not have that may lead to a higher level of well-being. Whenever someone interprets their needs and wants as to being satisfied or not, they then develop their cognitive well-being.

Affective Well-Being

These are the different levels of affect on well-being: "high negative affect is represented by anxiety and hostility; low negative affect is represented by calmness and relaxation; high positive affect is represented by a state of pleasant arousal enthusiasm and low positive affect is represented by a state of unpleasantness and low arousal (dull, sluggish)". Well-being is most usefully thought of as the dynamic process that gives people a sense of how their lives are going, through the interaction between their circumstances, activities and psychological resources or 'mental capital' or "You may say that it is a state of complete wellness.

Psychology in Well-Being

The correlation between well-being and positive psychology has been proven by many social scientists to be strong and positive one. According to McNulty (2012) "positive psychology at the subjective level is about valued subjective experiences". Well-being is an important factor in this subjective experience, as well as, contentment, satisfaction of the past, optimism for the future and happiness in the present. People are more likely to experience positive psychology if they take in the good things in each experience or situation. Even in the past if a person only focuses on the negative the brain will only be able to recognise the negative. The more the brain has access to the negative the easier, it becomes because that is what is more memorable. It takes more effort for the brain to remember the positive experiences because typically it is the more smaller actions and experiences that are the positive ones. James McNulty (2012) research looks at this idea a little bit closer. She argues that, "well-being is not determined solely by people's psychological characteristics but instead is determined jointly by the interplay between those characteristics and qualities of peoples social environments". When people have well-being they are experiencing a sense of emotional Freedom. There is nothing negative that is holding them back from experiencing positive emotions. This is true if a person is in a certain setting because it has been proved in a past research that a certain setting can hold a lot of memories for an individual just because of what was shared there and the meaning of it. For this "well-being is often equated with the experience of pleasure and the absence of pain over time". The less psychological pain an individual is experiencing them more he or she is going to experience well-being.

When someone is positively well-being they are also experiencing a few other things. It involves a sense of self-fulfillment, which is the feeling of being happy and satisfied because one is doing something that fully uses your abilities and talents (Merriam-Webster). The feeling of having a purpose in life and connection with others are also contributors to the idea of well-being. When people feel as though they have a purpose in the world they feel like they belong; they feel like they matter.

Education and Well-Being

When talking about the school system, the idea of well-being gets a little foggy. It is argued that school should only be about learning and education but kids learn so much about social skills and themselves in school. When a child feels like they belong they are more likely to perform better in school. As well as being taught an education, they have to learn how to believe in themselves and create a purpose for themselves. If well-being is established in kids at a young age then it is more likely to play a part in their life as they get older. John White (2013) looked at public schools in Britain now and in the past. In the past schools only focused on knowledge and education but now Britain has moved to more of a broader direction. They started a program called Every Child Matters initiative, that seeks to enhance children's well-being across the whole range of children's services.

Subjective Well-Being

Subjective well-being is "based on the idea that how each person thinks and feels about his or her life is important". This idea is developed specifically in a person's culture. People base their own well-being in relation to their environment and the lives of others around them. Well-being is also subjective to how one feels other people in their environment view them, whether that be in a positive or negative view. Well-being is also subjective to pleasure and whether or not basic human needs are fulfilled, although one's needs and wants are never fully satisfied. The quality of life of an individual and a society is dependent on the amount of happiness and pleasure, as well as human health. Whether or not other cultures is subjective to their culture is based on what kind of culture it is. "Collectivistic cultures are more likely to use norms and the social appraisals of others in evaluating their subjective well-being, whereas those individualistic societies are more likely to heavily weight the internal frame of reference arising from one's own happiness".

Ethnic Identity and Well-Being

Ethnic identity plays a crucial role in someone's cognitive well-being. Studies show that "both social psychological and developmental perspectives suggest that a strong, secure ethnic identity makes a positive contribution to cognitive well-being". Those in an acculturated society are able to feel more equal as a human being within their culture, therefore having a better well-being. This is also a crucial aspect when adapting to a new society.

Individual Roles and Well-Being

Individual roles play a part in cognitive well-being. Not only does having social ties improve cognitive well-being, it also improves psychological health. Having multiple identities and roles helps individuals to relate to their society and provide the opportunity for them to contribute more as they increase their roles, therefore creating a better cognitive well-being. Each individual role is ranked internally within a hierarchy of salience. Salience is "the subjective importance that a person attaches to each identity". Different roles an individual has have a different guidance to their well-being. Within this hierarchy, higher roles offer more of a source to their well-being and define more meaningfulness to their overall role as a human being.

Sports and Well-Being

According to (Bloodworth and Colleagues, 2012) sports and physical activities is a key contributor to the development of people's well-being. Sports being such a big influence on well-being it is conceptualized within a framework. These frameworks include impermanence, its hedonistic shallowness and its epistemological inadequacy. Arguments arise from these that the value of sports needs to be argued so humans can flourish. There can be problems from researching sports effecting well-being because some societies are not able to play sports. This is a deficiency in studying this sort of phenomenon.

Discrimination Based on Skin Colour

Discrimination based on skin colour, or colourism, is a form of prejudice or discrimination in which human beings are treated differently based on the social meanings attached to skin colour.

Colourism, a term coined by Alice Walker in 1982, is not a synonym of *racism*. "Race" depends on multiple factors (including ancestry);

therefore, racial categorization does not solely rely on skin colour. Skin colour is only one mechanism used to assign individuals to a racial category, but race is the set of beliefs and assumptions assigned to that category. Racism is the dependence of social status on the social meaning attached to race; colourism is the dependence of social status on skin colour *alone*. In order for a form of discrimination to be considered colourism, differential treatment must not result from racial categorization, but from the social values associated with skin colour.

Colourism can be found specifically in parts of Africa, Southeast Asia, East Asia, India, Latin America, and the United States. The abundance of colourism is a result of the global prevalence of "pigmentocracy," a term recently adopted by social scientists to describe societies in which wealth and social status are determined by skin colour. Throughout the numerous pigmentocracies across the world, the lightest-skinned peoples have the highest social status, followed by the brown-skinned, and finally the black-skinned who are at the bottom of the social hierarchy. This form of prejudice often results in reduced opportunities for those who are discriminated against on the basis of skin colour.

Africa

Skin bleaching is popular in Senegal and all across West Africa, especially among women.

Liberia

In Liberia, descendants of African-American settlers (renamed Americo-Liberians) in part defined social class and standing by raising people with lighter skin above those with dark skin. The first Americo-Liberian presidents such as Joseph Jenkins Roberts, James Spriggs Payne, and Alfred Francis Russell had considerable proportions of European ancestry. Most may have been only one-quarter or one-eighth African American. Other aspects of their rising to power, however, likely related to their chances for having obtained education and work that provided good livings.

In addition to rivalries among descendants of African Americans, the Americans held themselves above the native Africans in Liberia. Thus, descendants of Americans held and kept power out of proportion to their representation in the population of the entire country, so there was a larger issue than colour at work.

South Africa

Coloured people consist of three mixed race populations in South Africa who were given more social privilege than other, unmixed, indigenous African groups. During the apartheid era, in order to keep divisions and maintain a race-focused society, the government used the term *Coloured* to describe one of the four main racial groups identified by law: Blacks, Whites, Coloureds and Indians. (All four terms were capitalised in apartheid-era law.) Many Griqua began to self-identify as "Coloureds" during the apartheid era. There were certain advantages in becoming classified as "Coloured". For example, Coloureds did not have to carry a *dompas* (an identity document designed to limit the movements of the non-white populace), while the Griqua, who were seen as another indigenous African group, did.

Sudan

A popular phrase in Sudan is *al-Husnu ahmar* (beauty is red). Whiteness is the ideal colour in most Arab societies. The second ranking is *asmar* (light tan), followed by *dhahabi* (golden), *gamhi* (wheatish), *khamri* (the colour of wine), *akhdar* (light black/green). *Akhdhar* is used as a polite alternative to 'black' in describing the colour of a dark-skinned Arab. Last and least is *azraq*, literally "blue", used interchangeably with *aswad* to mean "black" — although in the past it referred to whiteness or light skin.

Asia

In Asia a preference for lighter skin remains prevalent and skin whitening cosmetic products are popular. Four out of ten women surveyed in Hong Kong, Malaysia, the Philippines and South Korea used a skin whitening cream, and more than 60 companies globally compete for Asia's estimated $18 billion market.

Skin whitening in Asia has a long history and stems back to Ancient China and Japan. To these cultures to be light in an environment where the sun is harsh, meant that one was upper class and rich enough to stay indoors and leave the outdoor work to servants. In many Asian cultures colourism is taught to children in the form of fairy tales, just like the Grimm fairy tales with the light skinned white princess or maiden, Asian mythological protagonists are typically fair and depict virtue, purity, and goodness. Lightness is equated with feminine beauty, racial superiority, and power and has strong influences on marital prospects, employment, status, and income.

With globalization, the obsession with whiteness has become even more prevalent in Asia's newfound hypercommercialism and consumer culture. In metropolises such as Hong Kong "get white" messages are inescapable, with light skinned Asian models covering billboards, magazine covers and counter spaces at the department stores. Some two thirds of men surveyed in Hong Kong said they prefer lighter skinned women and almost half of young Asian adults have admitted to use skin whiteners in attempts to achieve a lighter complexion. There have been issues of mercury poisoning in consumers of certain bleaching products in Asia, with some products reported to contain up to 60,000 times the acceptable dose.

India

Individuals in India have a tendency to see whiter skin as more beautiful. This can be traced to the invasion of Muslim emperors who were mostly light skinned. They referred to two classes of people, the white-skinned Muslims and the black-skinned Indian. The Aryans were religious and followed the Vedas, performing all the rituals while the Dasas (at a later stage) merged into the Shudra. The major God of Hinduism, Lord Krishna, the Sanskrit word for "Black" or "Dark", was the charioteer of Arjuna, both of whom were dark coloured. But sometimes Krishna is also translated as "all attractive"

Almost all the Goddesses in every art form are depicted as beautiful, sweet, charming and light skinned except Goddess Kali who has a dark skin and is described as 'the mother of all'.

The discrimination based on skin colour was most visible in British India, where skin colour served as a signal of high status for the foreign British. Thus, those individuals with a lighter skin colour enjoyed more privileges, were considered to have a more affluent status and gained preference in education and employment. Darker skinned individuals were socially and economically disadvantaged due to their skin tone. The caste system in India too involves complications of skin colour. British historians claimed that since the upper castes were not involved in tedious labour and weren't as exposed to the sun as the lower castes, they use to stay indoors and thus possessed lighter skin. The lower castes on other hand had higher melanin concentration in their skin cells due to continued exposure to sun from working in agricultural fields and outdoors.

Stereotypes Prevalent in the Society

Children are complimented by relatives and friends for being the 'fairer one', in teenage and this bias keeps growing with age. It can be blamed on peer pressure, societal prejudices or irresponsible advertising and product manufacturing.

Bleaching Creams

Skin-whitening cosmetics, popularized by Dutch company HUL are a multi-billion dollar industry pushing the idea that beauty equates with white skin and that lightening dark skin is both achievable and preferable. In a country such as India, with issues such as employment and relationships often resting on skin tone, people invest in skin-whitening creams in the hope of a better existence. Capitalizing on this inequality, hundreds of products are peddled by corporations, among them armpit whitener, genital whitener and fairness baby oil. Nearly all major cosmetic companies (like Dove, Nivea, Pond's, Garnier, Neutrogena, Olay) sell products that claim alter genes to suppress melanin.

Matrimonial Advertisements

Skin colour preference in matrimonial matters is something certainly not unique to India; however the way it gets expressed is most certainly distinctive from that in any other society.

Example: I am a TV journalist and working in no.1 news channel of India. I want to be the most dynamic and popular news anchor of India. I want to earn a lot of money in my life through my occupation. My wife 'must be very fair' and beautiful, who can understand and respect my feelings.

White-Collar Jobs Only for the White-Skinned?

The deep-rooted colour bias has ensured that in certain professions such as aviation, films and many other white collar jobs, people with light skin are generally preferred. In the western state of Maharashtra in India, about 100 tribal girls, who were trained to be airhostesses and cabin crew under a government scholarship programme aimed at empowering them, were denied jobs apparently because of their darker skin colour. Only eight of them landed jobs, but just as ground staff.

Indian fashion industry which is based on Western fashion is having fixation with 'fair' models is only ¬getting stronger. In an audition for a fashion week, out of the 11 short listed models, six were foreigners. An Indian model confessed that white skin has almost become a prerequisite. Some of the selected models did not even match the minimum height ¬requirement of 5'8".

'Skin' Crayon

Hindustan Pencils, the manufactures of the popular Nataraj and Apsara pencils have started a Colorama crayon series which has a peach-coloured crayon labelled as 'skin', even though it is clearly not the skin tone of most Indians. In a country with as many skin tones as India, labelling one particular shade as 'skin' colour and that shade in turn being used to represent skin in all human caricatures unknowingly deepens the colour bias against skin tone at a very tender age. This is not unique to India. Companies like Crayola, Faber-Castell and Camlin have crayons labelled as 'flesh' but Crayola chose to rename its 'flesh' crayon as 'peach' in 1962 in response to the US civil rights movement. It also introduced a special set of eight 'Multicultural Crayons' representing different skin tones.

Facebook-Photo Editing and Sharing

Hindustan Unilever, the manufacturer of Fair&Lovely, under its cosmetic brand name Vaseline, recently launched an application to make the skin of Facebook users look lighter in their profile pictures. There also exists a widespread practice of using photo editing software to make one self look lighter in photographs.

Campaigns and Petitions

Bollywood filmmaker Shekhar Kapur, who directed films such as "Bandit Queen" and "Elizabeth", started a campaign with the Twitter hash tag "adswedontbuy" to protest against irresponsible ads, including ads for skin whitening creams. Millions joined the discussion within a period of 24 hours.

'Women of Worth' is an organisation working towards empowerment of women. It started an awareness campaign called 'Dark is Beautiful' that seeks to draw attention to the unjust effects of skin colour bias and also celebrates the beauty and diversity of all skin tone. In January, they delivered a petition of 30,000 signatures to cosmetic company Emami, calling on them to withdraw a particularly discriminatory advert for Fair and Handsome. On this Emami's managing director said, "there is a need in our society for fairness creams, so we are meeting that need." He refused to withdraw the ad. Undeterred, Dark is Beautiful is lobbying the Advertising Council of India to legislate against adverts that discriminate against dark skin.

A second-year law student at Bangalore's National Law School filed a complaint against Hindustan Pencils at the district-level consumer forum in Bangalore in June in relation to the 'skin' crayon, accusing the company of being racist for promoting the idea that there is only one kind of acceptable skin colour – peach, in a country where most people have darker skin in varying tones of brown. When he lost the case at the district forum in October 2013, he took it up to the State Consumer Commission, where it is now being heard. He has asked for the removal of the label 'skin' of the crayon along with compensation of Rs 100,000 for hurting his sentiments. With 10 other people, he has started an NGO called Brown n' Proud which aims to raise awareness about 'rangbhed', or colour discrimination.

The Beginning of Change

Many Bollywood actors have showed support for a change in attitude against skin colour discrimination. Actor and activist Nandita Das became the face of the Dark is Beautiful campaign and also believes that the name of a crayon is important, even though it may seem like a small issue. "After all, children are sponges and absorb more than we think they do." An entire segment in Madhur Bhandarkar's Traffic Signal is devoted to an anti lightness-cream rant. The category's ads has been pilloried in global media for promoting a kind of "racism".

Many people now feel that they do not have to get light skin to succeed in life. Brown or dark skin is being embraced by young consumers which was reflected in the declines of the sales of Fair & Lovely skin lightening products (the largest player with nearly 60% of the market share) by 4.2% in 2013 while the sales of Fair & Handsome dipped by 14%.

The Advertising Standards Council of India, a self-regulatory body has proposed a draft of new guidelines that tells lightness advertisements not to show darkerskinned people as unhappy, depressed, or disadvantaged in any way by skin tone, and should not associate skin colour with any particular socio-economic class, ethnicity or community. According to Sam Balsara, chairman and managing director, Madison World and a former chairman of ASCI, "The reason for these guidelines is to make it clear to advertisers as to what society finds acceptable and what it doesn't." Though the Congress party led by Italy born Sonia Gandhi has done little to implement it.

Latin America

Brazil has the largest population of African descendants (living outside of Africa) in the world. This large number was a result of the African Slave trade. In Brazil, skin colour plays a large role in differences among the races. Individuals with lighter skin and who are racially mixed generally have higher rates of social mobility.

There are a disproportionate number of mostly European descent elites than those of visible African descent. There are large health, education and income disparities between the races in Brazil.

In parts of Latin America, light skin is seen as more attractive. In Mexico and in Brazil, light skin represents power. A dark skinned person is more likely to be discriminated against in Brazil. Most Latin American (particularly South American) actors and actresses have mostly European features - blond hair, light or light-mixed eyes, protruding narrow noses, straight hair and/or pale skin; the same situation happens in Hispanic media of the United States. A light skinned person is considered to be more privileged and have a higher social status. A person with light skin is considered beautiful and it means that the person has more wealth. Those with dark skin and frizzy hair tend to be among the region's poorest and most disenfranchised.

United States

Within the United States, colourism can be observed among all races. Although it occurs most notably among African Americans, it also occurs among Latinos, Asian Americans, Indian Americans, Native Americans, and even among European Americans.

African Americans

History: European colonialism created a system of white supremacy and racist ideology, which led to a structure of domination that privileged whiteness over blackness. Biological differences in skin colour were used as a justification for the enslavement and oppression of Africans, developing a social hierarchy that placed whites at the top and blacks at the bottom. The desire to rise out of this lower position ultimately caused internalized divisions among African Americans.

Miscegenation, the mixing of different racial groups (commonly through the sexual exploitation of black female slaves by white male slave owners and, after emancipation, black female prostitutes exchanging sex for money with white male customers) resulted in a large number of mixed race individuals with both African and European ancestry. Terminology was also developed to distinguish various levels of African ancestry. The terms mulatto, quadroon and octoroon were used to identify a black person with one-half, one-fourth and one-eighth of African ancestry, respectively. Slaves with lighter complexion were allowed to engage in less strenuous tasks, like domestic duties, while the darker slaves participated in hard labour, which was more than likely outdoors. A partial white heritage also gave light-skinned blacks more economic value and caused them to be viewed as smarter and superior to dark-skinned blacks, allowing more advantages in a white-dominated society, such as broader opportunities for education and the acquisition of land and property.

To prevent any confusion in regard to racial classification and to prohibit blacks with white ancestry from gaining the same legal status as full-blooded whites, the rule of hypo-descent, or the "one-drop rule" was mandated. According to the "one-drop rule," even the smallest amount of African ancestry (or a drop of African blood) legally defined a person as black. After the abolition of slavery in 1865, however, colourism created an internalized structure of hierarchy and division within the black community, as lighter-skinned blacks began to set themselves apart by socializing, marrying and procreating with one another. Around the beginning of the twentieth century, separatist standards, such as the brown paper bag, comb, pencil, and flashlight tests began to be implemented. Also, exclusionary social clubs and societies were developed to create colour divisions within black America that would shape socially constructed ideas about skin colour.

Examples of African American Colourism

Brown Paper Bag Test: The phrase "brown paper bag test" has traditionally been used by African Americans throughout the 20th and 21st century with reference to a ritual once practiced by certain African-American sororities and fraternities who would not let anyone into the group whose skin tone was darker than a paper bag. Also known as a paper bag party, these lighter-skinned social circles reflected an idea of exclusion and exclusiveness. The notion of the "paper bag" has carried a complex and obscure meaning in black communities for many decades. The reason for the usage of the "paper bag" is because the colour of the paper bag is considered to be the "centre" marker of blackness that distinguishes "light skin" from "dark skin" on a continuum stretching infinitely from black to white. Also, the brown paper bag is believed to act as a benchmark for certain levels of acceptance and inclusion. Spike Lee's film *School Daze* satirized this practice at historically black colleges and universities. Along with the "paper bag test," guidelines for acceptance among the lighter ranks included the "comb test" and "pencil test," which tested the coarseness of one's hair, and the "flashlight test," which tested a person's profile to make sure their features measured up or were close enough to those of the Caucasian race.

The Bleaching Syndrome

A phenomenon known as "The bleaching syndrome," constructed by Dr. Ronald E. Hall at Michigan State University in the early 1990s which refers to the process of attempting to lighten one's skin, has made a significant impact on the commercial industry and the lives of African Americans since the beginning of the 21st century. The roots of the skin bleaching phenomenon stem from African Americans internalizing dominant cultural ideas, without the possibility of full assimilation into American society. A psychological conflict is thus created, causing African Americans to develop a disdain for dark skin as it counters dominant cultural ideals. In an attempt to simultaneously reduce this conflict and enable assimilation, many African Americans developed the bleaching syndrome. Since the degree of assimilation correlates with skin colour based on dominant cultural standards, light skin is crucial relative to the degree of assimilation in the United States. Coincidentally, lighter skin is thought to be an ideal point of reference for attractiveness and marital partner selection among African Americans. The practice of applying light skin as a point of reference, however, is considered to be culturally self-destructive. Ultimately, the bleaching syndrome is a manifestation of the conflicting circumstances regarding the assimilation of African Americans into dominant cultural values.

Skin Colour Paradox

The skin colour paradox refers to the fact that no matter how differently African Americans are treated based on their skin colour, their political and cultural attitudes about "blackness" as a form of identity and their feelings of relatedness and solidarity with other blacks tend to remain consistent. Although light-skinned African Americans receive many socio-economic advantages over dark-skinned African Americans, who have much more punitive relationships with the criminal justice system and greatly diminished prestige, and although African Americans are aware of this disparity in treatment and status, both light-skinned and dark-skinned African Americans have similar political attitudes towards discrimination and race solidarity.

Political scientists would suggest that skin colour is a characteristic perhaps equally important as religion, income, and education, which is why this paradox is so surprising, but studies show that skin colour has no real bearing on actual political preference. Affirmative action is another example of the paradox between colourism on the one hand and political preference on the other. Studies show that most African Americans that benefit from affirmative action come from families that are better educated and more well off, and historically this means that the lighter-skinned portion of the black group is receiving the majority of the aid. Yet beneficiaries of this special treatment tend to hold on to their political identification with "blackness."

Stereotypes

The light to dark hierarchy within the African American race is one that has existed since the time of slavery, but its problems and consequences are still very evident and lead to various stereotypes. Darker skinned blacks are more likely to have negative relationships with the police, less likely to have higher education or income levels, and less likely to hold public office. Darker skinned people are also considered less intelligent, less desirable in women mostly, and are overall seen as inferior to lighter-skinned people. Studies have shown that when measuring education and family income, there is a positive sloping curve as the skin of families gets lighter. This does not prove that darker skinned people are discriminated against, but it provides insight as to why these statistics are recurring. Lighter skinned people tend to have higher social standing, more positive social networks, and more opportunities to succeed than those of a darker persuasion. Scientists believe this advantage is due to not only to their ancestors' benefits, but also to skin colour. In criminal sentencing, medium to dark-skinned African Americans are likely to receive sentences 2.6 years longer than those of whites or light-skinned African Americans, and when a white victim is involved, those with more "black" features are likely to receive a much more severe punishment, reinforcing the idea that those of lighter complexion are of more "value."

The perception of beauty can be influenced by racial stereotypes about skin colour; the African American journalist Jill Nelson wrote that "to be both prettiest and black was impossible" and elaborated:

As a girl and young woman, hair, body, and colour were society's trinity in determining female beauty and identity, the cultural and value-laden gang of three that formed the boundaries and determined the extent of women's visibility, influence, and importance. For the most part, they still are. We learn as girls that in ways both subtle and obvious, personal and political, our value as females is largely determined by how we look. As we enter womanhood, the pervasive power of this trinity is demonstrated again and again in how we are treated by the men we meet, the men we work for, the men who wield power, how we treat each other and, most of all, ourselves. For black women, the domination of physical aspects of beauty in women's definition and value render us invisible, partially erased, or obsessed, sometimes for a lifetime, since most of us lack the major talismans of Western beauty. Black women find themselves involved in a lifelong effort to self-define in a culture that provides them no positive reflection.

Asian Americans

In 2003 researchers at the University of California Santa Barbara conducted a survey of 99 Asian Americans on the issue of colourism in their communities. The ethnicities surveyed included Filipino, Japanese, Cambodian, Korean, and Chinese individuals. The respondents answered a number of questions varying from skin colour preference, eye shape, face shape, etc. Certain themes emerged from the survey that indicated that colourism is also a part of the Asian American communities and the consensus was that light-skinned people were more beautiful than dark people. One theme that was consistent was that "beauty is light" and another was to "not be romantically linked with or marry dark people". Many respondents reported that a member in their family had openly discouraged dating a dark-skinned Asian American. Similar to African American communities, Asian American women are more likely to be held to the lighter skin standard than Asian American men. This gender difference is due to the fact that the culture, family, and ideologies are strongly based in patriarchy where a woman submits to the man's "ideal of beauty". The preference for lighter skin in Asian communities dates back to traditional Asian standards of beauty brought over by immigrants, however exposure to images of Western ideal beauty in the media has also reinforced Asian American women's desires for lighter skin. In African American and Latin American communities, various skin bleaching creams and treatments are offered in the Asian communities, promoted by advertisements featuring light, almost pale, skin toned Asian models.

Media and Public Perception

The media is responsible for influencing beliefs regarding ideas of beauty in the African American community. Mass media productions often perpetuate discrimination based on skin colour. African Americans possessing lighter skin complexion and "European features," such as lighter eyes, and smaller noses and lips have more opportunities in the media industry. For example, film producers hire lighter-skinned African Americans more often, television producers choose lighter skinned cast members, and magazine editors choose African American models that resemble European features. As a result, the media industry sends the messages that African Americans with Eurocentric features are more likely to be accepted, diminishing the status of darker-skinned African Americans.

In regards to the magazine industry, African American women are rarely showcased in the most popular magazines. Therefore, African American girls have difficultly identifying with the models showcased in these magazines, because they do not represent the type of women that they come into contact with in their own communities. There are also biases towards Caucasians in the advertisements used in these magazines. Recent studies have indicated that the number of racially biased advertisements in magazines have increased over the years. A content analysis conducted by Scott and Neptune (1997) shows that less than one percent of advertisements in major magazines featured African American models. When African Americans did appear in advertisements they were mainly portrayed as athletes, entertainers or unskilled laborers. In addition, seventy percent of the advertisements that features animal print included African American women. Animal print reinforces the stereotypes that African Americans are animalistic in nature, sexually active, less educated, have lower income, and extremely concerned with personal appearances.

Concerning African American males in the media, darker skinned men are more likely to be portrayed as violent or more threatening, influencing the public perception of African American men. Since dark-skinned males are more likely to be linked to crime and misconduct, many people develop preconceived notions about the characteristics of black men. Through extreme gangsta rap music, reality crime shows, and newscasts, crime has been defined by contemporary media and given a black face despite statistics that paint a different picture. For example, cocaine use has been found to be higher among whites, but African Americans are the dominant figures seen on crime shows such as *Cops*. However, John Langely, the creator of Cops has stated in an interview that he actually over-represents whites in order to deflect accusations of racism, and that if he didn't, blacks would be even more heavily represented on the show than they already are.

The negative public perception of darker-skinned African American places them at a disadvantage in other aspects of society, such as the workforce. Skin colour plays a significant role in the acceptance of African Americans in the workforce and can even hold more importance than an individual's credentials and ability. For example, hiring managers generally have a different perception of a light-skinned African American female applicant, compared to a dark-skinned female applicant. Light-skinned African American women were found to have higher salaries than dark-skinned women, and light-skinned women were more satisfied with their jobs in regards to pay and advancement opportunities. Researchers have also termed dark-skinned women as being in a "triple-jeopardy" situation because all three aspects of their identity—their gender, race and skin-tone—can have negative and harmful implications on occupational opportunities and overall feelings of competency.

Despite exclusion and bias, the media has made an attempt to correct some of the negative images of African Americans. Examples of shows in which African Americans have been positively portrayed in the past include *The Cosby Show*, *The Fresh Prince of Bel-Air*, and *A Different World*. In addition, new television specials such as *Black Girl's Rock* and *My Black is Beautiful* highlight African American men and women for their contributions to society. Overall, these media changes have helped to provide a unique and more accurate representation of black culture in the twenty-first century. Television networks such as *Centric* and *TV One* and magazines such as *Essence* and *Ebony* play a major role in portraying African Americans in a more positive light than they have been portrayed in the past.

Globalization

The issue of colourism has gone global especially with the rise of globalization and advancing technologies. Through globalization multinational media conglomerates export U.S. culture, products, and cultural imperialism, including cultural images of race and colour preference. The global media implications are that Eurocentric physical features which are less aligned with native ancestry are found to be more attractive. The U.S. exports images of the good life, white beauty and affluence which many people in other poorer countries yearn for. The Philippines is a good example of this due to the fact that the country is an intersection of internalized colonial values and a culture of new global viewing. Just as in other Third World countries, the Philippines contemporary culture valorizes American culture and places high value on white beauty. In Korea, women pay high sums of money for eyelid surgery and skin bleaching products to achieve Westernized white beauty. Plastic surgery and skin bleaching are both derived from the globalization of U.S. dominant views of fair beauty and images of lighter-skinned celebrities are widely valued in the global market place. This is the reason as to why colourism is so difficult to battle, because it is so pervasive with images supporting this system of inequality reaching all over the globe. The advancement of television, film, print ads, and internet featuring lighter skin as a cultural ideal influences Third World and modernising countries, informing them that to be of lighter skin colour is not only a way to live the "good life" but is also culturally imperative.

Skin Bleaching

Skin bleaching has long since been one of the oldest forms of achieving fair skin and has become a multibillion dollar industry. This obsession with whiteness has not faded over time; a survey concluded that three quarters of Malaysian men thought their partners would be more attractive if they had lighter skin complexions. Women in countries such as South Korea, India, China, Saudi Arabia, and Uganda, use toxic skin bleaching creams to achieve a lighter skin complexion. Especially with the increase in globalization, many post colonial and Third World nations have seen a rise in the purchasing of skin lightening creams to achieve a Eurocentric appearance. Skin bleachers, which are marketed as 'beauty products' go by many names: skin lighteners, skin whiteners, skin-toning creams, skin fading gels, etc., however, the promises are all the same: that the product will help to reduce melanin in its consumers. In countries throughout Asia, these skin bleachers can be purchased in malls, drug stores, and even on the internet. These markets thrive on the vulnerabilities, fears, and taps into the cultural beliefs of countries that believe light-skinned is more valued. In the India, the highest selling brand of skin lightening bleach cream, Fair & Lovely, promotes that the product produces dignity and that to be fair skinned is aspirational. There are many reasons for the increasing global phenomenon of skin lighteners, from one's skin being too dark, to attracting romantic prospects or to be popular and fashionable. However, these skin bleaching products usually contain three harmful ingredients: mercury, hydroquinone, and/or corticosteroids. All of these chemicals can be extremely dangerous and fatal, and most of the products are made outside the U.S. and Europe and are therefore less subject to strict regulation.

Tanning

In the 20th century there has been a shift towards a preference for darker, tanned skin in white communities. The beginning of this change has been attributed to Frenchwoman Coco Chanel making tanned skin seem fashionable, luxurious and healthy in Paris in the 1920s. Tanned skin has become associated with the increased leisure time and sportiness of wealth and social status while pale skin is associated with indoor office work. Several studies have found tanned skin is regarded as both more attractive and healthier than pale skin, and there is a direct correlation between the degree of tanning and perceived attractiveness especially in young women. White celebrities such as Heidi Klum, Victoria Beckham and Katy Perry have used artificial tanning to darken their natural skin tone and self-tanning has become the fastest growing sector in the cosmetics market worldwide. Artificial tanning methods can cause serious health issues, with a 75% increase in the risk of melanoma associated with tanning beds and the potential of inhaling or ingesting potentially harmful chemicals.

Chapter 5

Sexism and Ageism

Sexism

Sexism or gender discrimination is prejudice or discrimination based on a person's sex or gender. Sexist attitudes may stem from traditional stereotypes of gender roles, and may include the belief that a person of one sex is intrinsically superior to a person of the other. A job applicant may face discriminatory hiring practices, or (if hired) receive unequal compensation or treatment compared to that of their opposite-sex peers. Extreme sexism may foster sexual harassment, rape and other forms of sexual violence.

Origin of the Term

According to Fred R. Shapiro, in *American Speech* (Vol. 60, No. 1, Spring 1985), the term "sexism" was most likely coined on November 18, 1965, by Pauline M. Leet during the Student-Faculty Forum at Franklin and Marshall College. The term appears in Leet's forum contribution titled "Women and the Undergraduate", in which she defines it by comparing it to racism, saying in part, "When you argue...that since fewer women write good poetry this justifies their total exclusion, you are taking a position analogous to that of the racist — I might call you in this case a "sexist"... Both the racist and the sexist are acting as if all that has happened had never happened, and both of them are making decisions and coming to conclusions about someone's value by referring to factors which are in both cases irrelevant."

Also according to Shapiro, the first time the term "sexism" appeared in print was in Caroline Bird's speech "On Being Born Female", which was published on November 15, 1968, in *Vital Speeches of the Day*

(p. 6). In this speech she said in part, "There is recognition abroad that we are in many ways a sexist country. Sexism is judging people by their sex when sex doesn't matter. Sexism is intended to rhyme with racism. Both have been used to keep the powers that be in power."

Sheldon Vanauken is sometimes falsely claimed to have coined the term "sexism" in his pamphlet "Freedom for Movement Girls – Now". In the pamphlet the term "sexism" appears without any citation of Leet or Bird, thus leading to the false idea that Vanauken coined it himself.

History

Figure: *Sati, or self-immolation by widows was prevalent in Hindu society during the Middle Ages.*

Certain forms of sex discrimination are illegal in some countries; in others, discrimination may be legally sanctioned under a variety of circumstances.

Ancient World

According to Peter Stearns, women in pre-agricultural societies held equal positions with men; it was only after the adoption of agriculture and sedentary cultures that men began to institutionalize the concept that women were inferior to men. Definitive examples of sexism in the ancient world included written laws preventing women from participating in the political process; for example, Roman women could not vote or hold political office.

Witch Hunts and Trials

Throughout history, women accused of witchcraft have been targeted by religious and state authorities, subjected to violence, prosecuted, and executed. The witch trials in the early modern period were a period of witch hunts between the 15th and 18th centuries, when across early modern Europe and to some extent in the European colonies in North America, there were widespread claims that malevolent Satanic witches were operating as an organised threat to Christendom. Several authors argue that the widespread misogyny of that period played a role in the persecution of these women.

In *Malleus Malificarum*, the book which played a major role in the witch hunts and trials, the authors argue that women are more likely to practice witchcraft than men, and write that:

All wickedness is but little to the wickedness of a woman ... What else is woman but a foe to friendship, an inescapable punishment, a necessary evil, a natural temptation, a desirable calamity, a domestic danger, a delectable detriment, an evil of nature, painted with fair colours!

Today, practicing witchcraft remains illegal in several countries, including Saudi Arabia, where it is punishable by death; and in 2011 a woman was beheaded in that country for 'witchcraft and sorcery'.

Coverture and Other Marriage Regulations

Until the 20th century U.S. and English law observed the system of coverture, where "by marriage, the husband and wife are one person in law; that is the very being or legal existence of the woman is suspended during the marriage". U.S. women were not legally defined as "persons" until 1875 (*Minor v. Happersett*, 88 U.S. 162).

In France married women received the right to work without their husband's permission in 1965, and in West Germany women obtained this right in 1977 (women in East Germany enjoyed more rights). In Spain during the Franco era a married woman required her husband's consent (*permiso marital*) for nearly all economic activities, including employment, ownership of property and travelling away from home; the *permiso marital* was abolished in 1975.

In many countries, women still lose significant legal rights at marriage. For example, Yemeni marriage regulations state that a wife must obey her husband and must not leave home without his permission.

In Iraq husbands have a legal right to "punish" their wives. The criminal code states at Paragraph 41 that there is no crime if an act is committed while exercising a legal right; examples of legal rights include: "The punishment of a wife by her husband, the disciplining by parents and teachers of children under their authority within certain limits prescribed by law or by custom".

In the Democratic Republic of Congo the Family Code states that the husband is the head of the household; the wife owes her obedience to her husband; a wife has to live with her husband wherever he chooses to live; and wives must have their husbands' authorization to bring a case in court or to initiate other legal proceedings.

Gender Stereotypes

Gender stereotypes are widely held beliefs about the characteristics and behaviour of women and men. Empirical studies have found widely shared cultural beliefs that men are more socially valued and more competent than women in a number of activities. For example, Susan Fiske and her colleagues surveyed nine diverse samples from different regions of the United States. Fiske found that members of these samples, regardless of age, consistently rated the category of "men" higher than the category of "women" on a multidimensional scale of competence. Dustin B. Thoman and others (2008) hypothesize that "the socio-cultural salience of ability versus other components of the gender-math stereotype may impact women pursuing math". Through the experiment comparing the math outcomes of women under two various gender-math stereotype components, which are the ability of math and the effort on math respectively, Thoman and others found that women's math performance is more likely to be affected by the negative ability stereotype, which is influenced by socio-cultural beliefs in the United States, rather than the effort component. As a result of this experiment and the socio-cultural beliefs in the United States, Thoman and others concluded that individuals' academic outcomes can be affected by the gender-math stereotype component that is influenced by the socio-cultural beliefs (Thoman et al. 2008).

There are huge areal differences in attitudes towards appropriate gender roles. For example, in the *World Values Survey*, responders were asked if they thought that wage work should be restricted to only men in the case of shortage in jobs. While in Iceland the proportion that agreed was 3.6%, in Egypt it was 94.9%.

Some people believe a phenomenon known as stereotype threat can lower women's performance on mathematics tests, creating a self-fulfilling stereotype of women having inferior quantitative skills compared to men. Stereotypes can also affect self-assessment; studies found that specific stereotypes (e.g., women have lower mathematical abilities) affect women's and men's perceptions of those abilities, and men assess their own task ability higher than women who perform at the same level. These "biased self-assessments" have far-reaching effects, because they can shape men and women's educational and career decisions.

American Gender Stereotypes

Gender roles are behaviours and tasks which society associates with each gender. In the United States, men are typically expected to be stoic, decisive, direct, athletic, strong, driven, and brave; women are expected to be emotional, nurturing, affectionate, home-oriented and forgiving.

According to the OECD, women's labour-market behaviour "is influenced by learned cultural and social values that may be thought to discriminate against women (and sometimes against men) by stereotyping certain work and life styles as 'male' or 'female'". The organisation contends that women's educational choices "may be dictated, at least in part, by their expectations that types of employment opportunities are not available to them, as well as by gender stereotypes that are prevalent in society". Professional discrimination against women in the workplace continues. One study found that 50 percent of women in government jobs reported that they had experienced discrimination while performing a work-related task, and 20 percent of female government employees reported that they experienced gender discrimination at work. Women do not receive equal pay for equal work, and are less likely to be promoted.

In 1833, women working in factories earned one-quarter of what men earned; in 2007, women's median annual paychecks were $0.78 for every $1.00 earned by men. A study showed that women comprised 87% of workers in the child-care industry, and 86 percent of the health-aide industry. Others contest the wage gap, saying the difference in pay primarily reflects different career and work-hours choices by men and women rather than sexism.

A 2009 study found that being overweight harms women's career advancement, but presents no barrier for men. Overweight or obese women were significantly under-represented among company bosses; however, a significant proportion of male executives were overweight or obese. The author of the study stated that the results suggest that "the 'glass ceiling effect' on women's advancement may reflect not only general negative stereotypes about the competencies of women, but also weight bias that results in the application of stricter appearance standards to women. Overweight women are evaluated more negatively than overweight men. There is a tendency to hold women to harsher weight standards."

Transgender people also experience workplace discrimination and harassment. Unlike sex-based discrimination, the refusal to hire (or firing) a worker for their gender identity or expression is legal in most countries and U.S. states.

Gender discrimination is closely related to the patriarchy. The stereotypes, to some extent, are formed because of the patriarchy. Patriarchy is a social system in which males are the primary authority figures central to social organisation, occupying roles of political leadership, moral authority, and control of property, and where fathers hold authority over women and children. It implies the institutions of male rule and privilege, and entails female subordination.

In Language

Sexism in language exists when language helps to devalue members of a certain sex and thus fosters gender inequality, reinforcing ideas of gender-based dominance. What is now termed sexist language often promotes the idea of inherently male superiority, resulting in discrimination against women.

Sexism in language has arisen as a topic of concern as academics have pointed to the role that language plays in articulating consciousness, ordering our reality, encoding and transmitting cultural meanings and affecting socialization. Researchers have pointed to the semantic rule in operation in language of the male-as-norm. This results in sexism as the male becomes the standard and those who are not male are relegated to the inferior. Examples include the use of generic masculine words which refer to all humanity, like man, master, father, and brother. Other examples include the use of the singular masculine pronoun (he, his, him). Further examples point to terms ending in 'man' that may be performed by individuals of either sex, such as businessman, chairman, policeman. Finally, ordering of words also relegates women to an inferior position, in examples like "man and wife". Sexism in language is considered a form of indirect sexism, in that it is not always overt. Words and expressions that can be used to refer to men or women may be strongly associated with men or women (i.e., be associated with a gender stereotype, such as doctor or nurse, respectively) or be gender neutral.

Sexist and Gender-Neutral Language

Various feminist movements in the 20th century, from liberal feminism and radical feminism to standpoint feminism, postmodern feminism and queer theory have all considered language in their theorizing. Most of these theories have maintained a critical stance on language that call on a change in the way speakers use their language. One of the most common calls is for gender-neutral language. Many have called attention, however, to the fact that the English language isn't inherently sexist in its linguistic system, but rather the way it is used becomes sexist and gender-neutral language could thus be employed. At the same time, other oppose critiques of sexism in language with explanations that language is a descriptive, rather than prescriptive, and attempts to control it can be fruitless.

Sexism in Languages Other than English

Languages other than English are also charged with sexism in language. Romanic languages such as French and Spanish have been accused of sexist structures, in that the masculine form is the default form. For example, due to pressure from several feminist groups, the word 'mademoiselle', meaning miss was declared banished from the French language in 2012 by Prime Minister François Fillon. Current pressure calls for the use of the masculine plural pronoun as the default in a mixed-sex group to change. As to Spanish, Mexico's Ministry of the Interior published a guide on how to reduce the use of sexist language.

German speakers have also raised questions about how sexism intersects with grammar.

In Chinese, some writers have pointed to sexism inherent in the structure of written characters. For example, the character for man is linked to those for positive qualities like courage and effect while the character for wife is composed of a female part and a broom, considered of low worth.

Anthropological Linguistics and Gender-Specific Language

Unlike the Indo-European languages, gender-specific pronouns in many other languages are a relatively recent phenomenon (occurring during the early 20th century). Cultural revolution resulted from colonialism in many parts of the world, with attempts to modernise and westernize local languages with gender-specific and animate-inanimate pronouns. As a result, gender-neutral pronouns became gender-specific.

Gender-Specific Pejorative Terms

Gender-specific pejorative terms intimidate or harm another person because of their gender. Sexism can be expressed in language with negative gender-oriented implications, such as condescension. Other examples include obscene language. Some words are offencive to transgender people, including "tranny", "she-male", or "he-she". Intentional misgendering (assigning the wrong gender to someone) and the pronoun "it" are also considered pejorative.

Occupational Sexism

Occupational sexism refers to discriminatory practices, statements or actions, based on a person's sex, occurring in the workplace. One form of occupational sexism is wage discrimination.

In 2008 the Organisation for Economic Co-operation and Development (OECD) found that while female employment rates have expanded and gender employment and wage gaps have narrowed nearly everywhere, on average women still have 20% less chance to have a job and are paid 17% less than men. The report stated:

In many countries, labour market discrimination – i.e. the unequal treatment of equally productive individuals only because they belong to a specific group – is still a crucial factor inflating disparities in employment and the quality of job opportunities... Evidence presented in this edition of the *Employment Outlook* suggests that about 8% of the variation in gender employment gaps and 30% of the variation in gender wage gaps across OECD countries can be explained by discriminatory practices in the labour market.

It also found that despite the fact that almost all OECD countries, including the U.S., have established anti-discrimination laws, these laws are difficult to enforce.

Tokenism

Research shows that women who enter predominantly male work groups often "experience the negative consequences of tokenism: performance pressures, social isolation, and role encapsulation". However, attributing these negative consequences to tokenism itself might camouflage their root cause: sexism, "and its manifestations in higher-status men's attempts to preserve their advantage in the workplace". Furthermore, there has been no causal link proven between the number of women working in an organisation/company and the improvement of their conditions of employment (even worse, having an increasing number of women in a work place without appropriately addressing the "sexist attitudes imbedded in male-dominated organisations, may exacerbate women's occupational problems").

Wage Gap

Several studies have found that women earn a smaller average wage than men. Many economists and feminist scholars have argued that this is the result of systemic gender-based discrimination in the workplace. Others, however, maintain that the wage gap is a result of differences between the choices that men and women make in the workplace, such as more women than men choosing to be full-time parents or work fewer hours to be part-time parents.

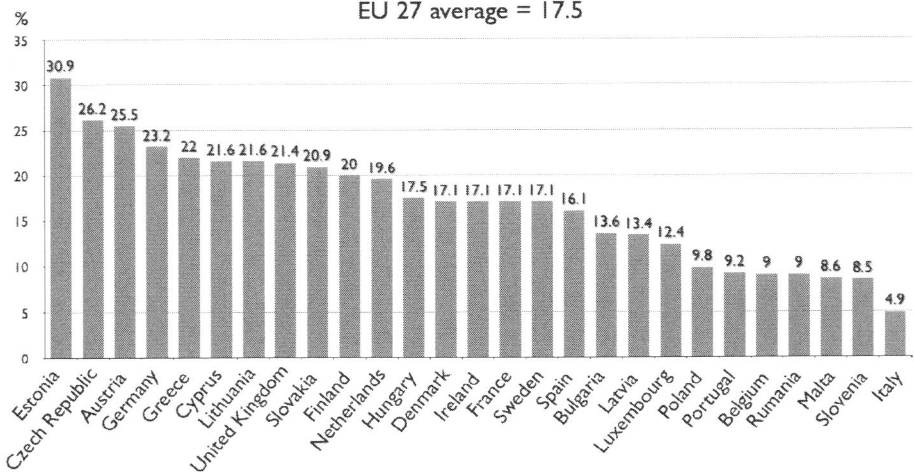

Figure: *Gender pay gap in average gross hourly earnings according to Eurostat 2008*

Eurostat found a persistent, average gender pay gap of 17.5 percent in the 27 EU member states in 2008. Similarly, the OECD found that female full-time employees earned 17 percent less than their male counterparts in OECD countries in 2009.

In the United States, the female-to-male earnings ratio was 0.77 in 2009; female full-time, year-round (FTYR) workers earned 77 percent as much as male FTYR workers. Women's earnings relative to men's fell from 1960 to 1980 (60.7 percent to 60.2 percent), rose rapidly from 1980 to 1990 (60.2 to 71.6 percent), leveled off from 1990 to 2000 (71.6 to 73.7 percent) and rose from 2000 to 2009 (73.7 to 77.0 percent). When the first Equal Pay Act was passed in 1963, female full-time workers earned 58.9 percent as much as male full-time workers. Research conducted in the Czech and Slovak Republics shows that, even after the governments have passed anti-discrimination legislation, two thirds of the gender gap in wages remained unexplained and segregation continued to "represent a major source of the gap". The gender gap can also vary across-occupation and within occupation. In Taiwan, for example, studies show how the bulk of gender wage discrepancies occur within-occupation. In Russia, research shows that the gender wage gap is distributed unevenly across income levels, and that it mainly occurs at the lower end of income distribution. Interestingly though, the research also finds that "wage arrears and payment in-kind attenuated wage discrimination, particularly amongst the lowest paid workers, suggesting that Russian enterprise managers assigned lowest importance to equity considerations when allocating these forms of payment".

The gender pay gap has been attributed to differences in personal and workplace characteristics between women and men (such as education, hours worked and occupation), innate behavioural and biological differences between men and women and discrimination in the labour market (such as gender stereotypes and customer and employer bias). Women interrupt their careers to take on child-rearing responsibilities more frequently than men. In Korea, for example, it has been a long-established practice to lay-off female employers upon marriage. A study by professor Linda Babcock in her book *Women Don't Ask* shows that men are eight times more likely to ask for a pay raise, suggesting that pay inequality may be partly a result of behavioural differences between the sexes.

However, studies generally find that a portion of the gender pay gap remains unexplained after accounting for factors assumed to influence earnings; the unexplained portion of the wage gap is attributed to gender discrimination. Estimates of the discriminatory component of the gender pay gap vary. The OECD estimated that approximately 30% of the gender pay gap across OECD countries is due to discrimination. Australian research shows that discrimination accounts for approximately 60 percent of the wage differential between women and men. Studies examining the gender pay gap in the United States show that a large portion of the wage differential remains unexplained, after controlling for factors affecting pay. One study of college graduates found that the portion of the pay gap unexplained after all other factors are taken into account is five percent one year after graduating and twelve percent a decade after graduation. A study by the American Association of University Women (AAUW) found that women graduates are paid less than men doing the same work and majoring in the same field.

Wage discrimination is theorized as contradicting the economic concept of supply and demand, which states that if a good or service (in this case, labour) is in demand and has value it will find its price in the market. If a worker offered equal value for less pay, supply and demand would indicate a greater demand for lower-paid workers. If a business hired lower-wage workers for the same work, it would lower its costs and enjoy a competitive advantage. According to supply and demand, if women offered equal value demand (and wages) should rise since they offer a better price (lower wages) for their service than men do.

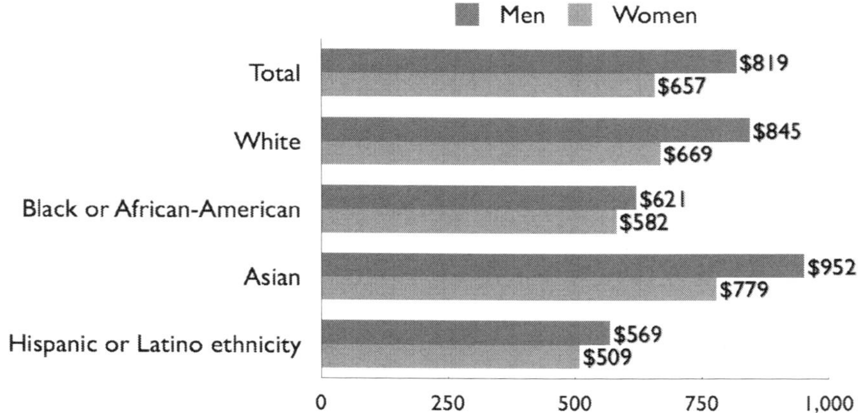

Figure: *Median weekly earnings of full-time wage and salary workers, by sex, race, and ethnicity, U.S., 2009.*

Research at Cornell University and elsewhere indicates that mothers are less likely to be hired than equally-qualified fathers and, if hired, receive a lower salary than male applicants with children. The OECD found that "a significant impact of children on women's pay is generally found in the United Kingdom and the United States". Fathers earn $7,500 more, on average, than men without children do.

Possible Causes for Wage Discrimination

According to Denise Venable at the National Centre for Policy Analysis, the "wage gap" is not the result of discrimination but of differences in lifestyle choices. Venable's report found that women are less likely than men to sacrifice personal happiness for increases in income or to choose full-time work. She found that among adults working between one and thirty-five hours a week and part-time workers who have never been married, women earn more than men. Venable also found that among people aged 27 to 33 who have never had a child, women's earnings approach 98% of men's and "women who hold positions and have skills and experience similar to those of men face wage disparities of less than 10 percent, and many are within a couple of points". Venable concluded that women and men with equal skills and opportunities in the same positions face little or no wage discrimination: "Claims of unequal pay almost always involve comparing apples and oranges".

There is considerable agreement that gender wage discrimination exists, however, when it comes to estimating its magnitude, significant discrepancies are visible. A meta-regression analysis concludes that "the estimated gender gap has been steadily declining" and that the wage rate calculation is proven to be crucial in estimating the wage gap. The analysis further notes that excluding experience and failing to correct for selection bias from analysis might also bring to incorrect conclusions.

Glass Ceiling Effect

"The popular notion of glass ceiling effects implies that gender (or other) disadvantages are stronger at the top of the hierarchy than at lower levels and that these disadvantages become worse later in a person's career."

In the United States, women account for 47 percent of the overall labour force, and yet they make up only 6 percent of corporate CEOs and top executives. Some researchers see the root cause of this situation in the tacit discrimination based on gender, conducted by current top executives and corporate directors (primarily male), as well as "the historic absence of women in top positions", which "may lead to hysteresis, preventing women from accessing powerful, male-dominated professional networks, or same-sex mentors". The glass ceiling effect is noted as being especially persistent for women of colour (according to a report, "women of colour perceive a 'concrete ceiling' and not simply a glass ceiling).

In the economics profession, it has been observed (Singell et all, 1996) that women are more inclined than men to dedicate their time to teaching and service. Since continuous research work is crucial for promotion, "the cumulative effect of small, contemporaneous differences in research orientation could generate the observed significant gender difference in promotion". In the hightech industry, research shows that, regardless of the intra-firm changes, "extra-organisational pressures will likely contribute to continued gender stratification as firms upgrade, leading to the potential masculinization of skilled hightech work".

The United Nations asserts that "progress in bringing women into leadership and decision making positions around the world remains far too slow."

Potential Remedies

One research suggests that a possible remedy to the glass ceiling could be increasing the number of women on corporate boards, which could subsequently lead to increases in the number of women working in top management positions. The same research suggests that this could also result in a "feedback cycle in which the presence of more female managers increases the qualified pool of potential female board members (for the companies they manage, as well as other companies), leading to greater female board membership and then further increases in female executives".

Objectification

Objectification is treating a person, usually a woman, as an object. By being objectified, a person is denied agency. Sexual objectification is where a person is viewed primarily in terms of sexual appeal or as a source of sexual gratification. This is sometimes regarded as a form of sexism. Nussbaum has identified the seven features of treating a human as an object as the following:

1. instrumentality: treating the object as a tool for the objectifier's purposes
2. denial of autonomy: treating the object as lacking in autonomy and self-determination
3. inertness: - treating the object as lacking in agency
4. fungibility: - treating the object as interchangeable with other objects
5. violability: - treating the object as lacking in boundaries-integrity
6. ownership: - treating the object as something that is owned by another (can be bought or sold);
7. denial of subjectivity: treating the object as something whose experiences and feelings (if any) need not be taken into account.

According to objectification theory (Fredrickson & Robert, 1997) objectification can have important repercussions on women, particularly young women, as it can lead to mental disorders (depression, eating disorders, etc.).

Objectification can take place in a variety of areas, such as in advertising and pornography.

In Advertising

While advertising used to portray women in obvious stereotypical ways (e.g. as a housewife), women in today's advertisements are no longer solely confined to the house. However, advertising today nonetheless still stereotypes women, albeit in more subtle ways, including by sexually objectifying them. This is problematic because there appears to be a relationship between the manner in which women are portrayed in advertising and people's ideas about the role of women in society. Research has shown that gender role stereotyping in advertising is linked to negative attitudes towards women, as well as more acceptance of sexual aggression against women and rape myth acceptance. Furthermore, gender role stereotyping in advertisements may be injurious to women, as it is linked to negative body image and the development of eating disorders.

Traditionally, in Western countries, advertising was regulated under "obscenity" laws. These dealt with nudity and "indecent" poses, not with objectification itself.

Today, some countries (for example Norway and Denmark) have laws against sexual objectification in advertising. Nudity is not banned, and nude people can be used to advertise a product if they are relevant to the product advertised. Sol Olving, head of Norway's Kreativt Forum (an association of the country's top advertising agencies) explained, "You could have a naked person advertising shower gel or a cream, but not a woman in a bikini draped across a car".

Other countries continue to ban nudity (on traditional obscenity grounds), but also make explicit reference to sexual objectification, such as Israel's ban of billboards that "depicts sexual humiliation or abasement, or presents a human being as an object available for sexual use". (Art 214)

Pornography

Anti-pornography feminist Catharine MacKinnon argues that pornography contributes to sexism by objectifying women and portraying them in submissive roles. MacKinnon, along with Andrea Dworkin, argues that pornography reduces women to mere tools, and is a form of sex discrimination. The scholars highlight the link between objectification and pornography by stating:

"We define pornography as the graphic sexually explicit subordination of women through pictures and words that also includes (i) women are presented dehumanized as sexual objects, things, or commodities; or (ii) women are presented as sexual objects who enjoy humiliation or pain; or (iii) women are presented as sexual objects experiencing sexual pleasure in rape, incest or other sexual assault; or (iv) women are presented as sexual objects tied up, cut up or mutilated or bruised or physically hurt; or (v) women are presented in postures or positions of sexual submission, servility, or display; or (vi) women's body parts—including but not limited to vaginas, breasts, or buttocks — are exhibited such that women are reduced to those parts; or (vii) women are presented being penetrated by objects or animals; or (viii) women are presented in scenarios of degradation, humiliation, injury, torture, shown as filthy or inferior, bleeding, bruised, or hurt in a context that makes these conditions sexual." Robin Morgan and Catharine MacKinnon suggest that certain types of pornography also contribute to violence against women by eroticizing scenes in which women are dominated, coerced, humiliated or sexually assaulted.

Some people opposed to pornography, including MacKinnon, charge that the production of pornography entails physical, psychological and economic coercion of the women who perform and model in it. Opponents of pornography charge that it presents a distorted image of sexual relations and reinforces sexual myths; it shows women as continually available and willing to engage in sex at any time, with any men, on their terms, responding positively to any requests. They write:

> *Pornography affects people's belief in rape myths. So for example if a woman says "I didn't consent" and people have been viewing pornography, they believe rape myths and believe the woman did consent no matter what she said. That when she said no, she meant yes. When she said she didn't want to, that meant more beer. When she said she would prefer to go home, that means she's a lesbian who needs to be given a good corrective experience. Pornography promotes these rape myths and desensitises people to violence against women so that you need more violence to become sexually aroused if you're a pornography consumer. This is very well documented.*

Defenders of pornography and anti-censorship activists (including sex-positive feminists) argue that pornography does not seriously impact a mentally healthy individual, since the viewer can distinguish between pornography and reality. They contend that both sexes are objectified in pornography (particularly sadistic or masochistic pornography, in which men are objectified and sexually used by women).

Prostitution

Some people argue that prostitution is a form of exploitation of women, which is the result of discrimination of women (a lack of professional opportunities for women, poverty) and of patriarchal views on sexuality which stipulate that unwanted sex with a woman is acceptable, that men's desires must be satisfied at any cost, and that women exist to serve men sexually. They argue that the consent of the prostitute is seriously undermined by direct (pimps) or indirect (social circumstances) coercion. The European Women's Lobby condemned prostitution as "an intolerable form of male violence".

Carole Pateman Writes that: "Prostitution is the use of a woman's body by a man for his own satisfaction. There is no desire or satisfaction on the part of the prostitute. Prostitution is not mutual, pleasurable exchange of the use of bodies, but the unilateral use of a woman's body by a man in exchange for money".

Media Portrayals

Some feminist scholars believe that media portrayals of demographic groups can both maintain and disrupt attitudes and behaviours toward those groups. These images are often harmful, particularly to women and racial and ethnic minority groups. For example, a study of African American women found they feel that media portrayals of African American women often reinforce stereotypes of this group as overly sexual and idealize images of lighter-skinned, thinner African American women (images African American women describe as objectifying). In a recent analysis of images of Haitian women in the Associated Press photo archive from 1994 to 2009, several themes emerged emphasizing the "otherness" of Haitian women and characterizing them as victims in need of rescue.

Sexist Jokes

Sexist jokes can be a form of sexual objectification, which reduces the subject of the joke to an object (e.g., "What do you do when your dishwasher won't work?" Answer: "You hit her"). They not only objectify women or men, but can also condone violence or prejudice against men or women. "Sexist humor—the denigration of women through humor—for instance, trivializes sex discrimination under the veil of benign amusement, thus precluding challenges or opposition that nonhumorous sexist communication would likely incur." A study of 73 male undergraduate students by Ford (2007) found that "sexist humor can promote the behavioural expression of prejudice against women amongst sexist men". According to the study, when sexism is presented in a humorous manner it is viewed as tolerable and socially acceptable. "Disparagement of women through humor 'freed' sexist participants from having to conform to the more general and more restrictive norms regarding discrimination against women." One example of sexist language is that used as slogans on the back of the Wicked Campers company such as "If you've ever met a woman with crooked teeth, you've met a woman who has given Chuck Norris a blow job."

Gender Discrimination

Gender discrimination is discrimination on the basis of actual or perceived gender identity. "Gender identity" is defined to mean "the gender-related identity, appearance, or mannerisms or other gender-related characteristics of an individual, with or without regard to the individual's designated sex at birth". Gender discrimination is theoretically different from sexism. Whereas, sexism is prejudice based on biological sex, gender discrimination specifically addresses discrimination towards identity based orientations, including third gender, genderqueer, and other non-binary identified people. Banning discrimination on the basis of gender identity and expression has emerged as a subject of contestation in the American legal system. Legal interpretation of "sex" under Title VII of the United States' Employment Non Discrimination Act (ENDA; H.R. 1755/S. 815) has, since the ruling of Price Waterhouse, encompassed both gender and sex. Since first introduced in Congress since 1993 at the 103rd Congress, the Employment Non-Discrimination Act (ENDA), has taken different stances on the inclusion of gender discrimination under the term "sex". More recently, the stated purpose of the legislation is "to address the history and persistent, widespread pattern of discrimination, including unconstitutional discrimination, on the basis of sexual orientation and gender identity by private sector employers and local, State, and Federal Government employers," as well as to provide effective remedies for such discrimination. However, under current law Title VII does not expressly prohibit discrimination based on gender identity or transsexualism. The myriad forms of discrimination associated with gender non-conformity are therefore not expressly addressed nor protected under the ENDA. According to a recent report by the Congressional Research Service, "although the majority of federal courts to consider the issue have concluded that discrimination on the basis of gender identity is not sex discrimination, there have been several courts that have reached the opposite conclusion in the years since the Supreme Court's decision in Price Waterhouse". Hurst states, "Courts often confuse sex, gender and sexual orientation, and confuse them in a way that results in denying the rights not only of gays and lesbians, but also of those who do not present themselves or act in a manner traditionally expected of their sex". Scholars have suggested amending the EDNA to include these gender orientations. However, counter arguments question whether Title VII is general enough to include sexual orientation or gender identity discrimination in legal claims. Gender discrimination is therefore, a gender identity based discrimination, whose codification into American and other legal systems has remained contested.

Transgender Discrimination

Transgender discrimination is discrimination towards peoples whose gender identity differs from the social expectations of the biological sex they were born with. Forms of discrimination include but are not limited to identity documents not reflecting one's gender, sex-segregated public restrooms and other facilities, dress codes according to binary gender codes, and lack of access to and existence of appropriate health care services. In a recent adjudication, the Equal Employment Opportunity Commission (EEOC) concluded that discrimination against a transgendered individual is sex discrimination. The National Transgender Discrimination Survey, the most extensive survey of transgender discrimination, in collaboration with the National Black Justice Coalition recently showed that Black transgender people in the United States suffer "the combination of anti-transgender bias and persistent, structural and individual racism" and that "black transgender people live in extreme poverty that is more than twice the rate for transgender people of all races (15%), four times the general Black population rate 9% and over eight times the general US population rate (4%)". In another study conducted in collaboration with the League of United Latin American Citizens, Latino/a transgender people who were non-citizens were most vulnerable to harassment, abuse and violence. Peoples of colour subjected to transgender discrimination suffer intersecting structural and individual levels of discrimination.

Gender Discrimination in Politics

Gender has been used, at times, as a tool of discrimination against women in the political sphere. Indeed, Women's suffrage was not achieved until 1893, when New Zealand was the first country to grant women the right to vote. Saudi Arabia was the last country to grant women the right to vote in 2011. Some Western countries allowed women the right to vote only relatively recently: Swiss women gained the right to vote in federal elections in 1971, and Appenzell Innerrhoden became the last canton to grant women the right to vote on local issues (in 1991, when it was forced to do so by the Federal Supreme Court of Switzerland). French women were granted the right to vote in 1944. In Greece, women obtained the right to vote in 1952.

While almost every woman today has the right to vote, there is still progress to be made for women in politics. Indeed, studies have shown that in several democracies including Australia, Canada and the United States, women are still represented using deep-rooted gender stereotypes in the press In fact, multiple authors have shown that gender differences in the media are less evident today than they used to be in the 1980s, but are nonetheless still present. Certain issues (e.g. education) are likely to be linked with female candidates, while other issues (e.g. taxes) are likely to be linked with male candidates. In addition, there is more emphasis on female candidates' personal qualities, such as their appearance and their personality, as females are portrayed as "emotional" and "dependent". This is problematic because voters' views of the female and male candidates may be affected, as well as their views of women's role in the political sphere

Gender discrimination in politics can also be shown in the imbalance of law making power between men and women. Lanyan Chen asserts that men hold more political power than women, serving as the gatekeepers of policy making. It is possible that this leads to women's needs not being properly represented. In this sense, the inequality of law making power also causes the gender discrimination in politics. The ratio of women to men in legislatures is used as a measure of gender equality in the UN created Gender Inequality Index.

Examples

Sexism takes a number of forms, and is sometimes subtle or unconscious.

Child Marriage

A child marriage is a marriage where one or both spouses are under 18. Girls are disproportionately the most affected. Child marriages are most common in South Asia, the Middle East and Sub-Saharan Africa, but occur in other parts of the world, too. The practice of marrying young girls is rooted in patriarchal ideologies of control of female behaviour, and is also sustained by traditional practices such as dowry and bride price. Child marriage is strongly connected with the protection of female virginity. UNICEF states that:

"Marrying girls under 18 years old is rooted in gender discrimination, encouraging premature and continuous child bearing and giving preference to boys' education. Child marriage is also a strategy for economic survival as families marry off their daughters at an early age to reduce their economic burden."

Consequences of child marriage include restricted education and employment prospects, increased risk of domestic violence, child sexual abuse, pregnancy and birth complications, social isolation. Early and forced marriage are defined as forms of modern-day slavery by the International Labour Organisation.

According to the UN, the ten countries with the highest rates of child marriage are: Niger (75%), Chad and Central African Republic (68%), Bangladesh (66%), Guinea, Mozambique, Mali, Burkina Faso, South Sudan, and Malawi.

Controlling Women's Dressing Attire

Laws that dictate how women must dress are seen by many international human rights organisations, such as Amnesty International, as a form of gender discrimination. Amnesty International states that:

"Interpretations of religion, culture or tradition cannot justify imposing rules about dress on those who choose to dress differently. States should take measures to protect individuals from being coerced to dress in specific ways by family members, community or religious groups or leaders."

In many places, women who do not dress in socially and legally proscribed ways are often subjected to violence (for instance by the authorities, such as the religious police, by family members, or by the community).

Criminal Justice

Some studies in the United States have shown that for identical crimes, men are given harsher sentences than women. Controlling for arrest offence, criminal history and other pre-charge variables, sentences are over 60 percent heavier for men. Women are more likely to avoid charges entirely, and to avoid imprisonment if convicted. The gender disparity varies according to the nature of the case (for example, there is a smaller gender gap in fraud cases than in drug trafficking and firearms). This disparity occurs in U.S. federal courts, despite guidelines designed to avoid differential sentencing.

There are many reasons postulated for the gender disparity. One of the most common is the larger caregiving role of women. Other possible reasons include the "girlfriend theory" (whereby women are seen as tools of their boyfriends), the theory that female defendants are more likely to cooperate with authorities, and that women are often successful at turning their violent crime into victimhood by citing defences such as postpartum depression and battered wife syndrome. However, none of these theories can account for the total disparity, and sexism has also been suggested as an underlying cause.

The criminal justice system has also been accused of discriminating against women. Provocation is, in many common law countries, a partial defence to murder, which converts what would have been murder into manslaughter. It is meant to be applied when a person kills in the 'heat of passion' upon being 'provoked' by the behaviour of the victim. This defence has been criticized as being gendered, favouring men, due to it being used disproportionately in cases of adultery, and other domestic disputes when women are killed by partners. As a result of the defence being considered as exhibiting a strong gender bias, and being a form of legitimization of male violence against women and minimization of the harm caused by violence against women, it has been abolished or restricted in several jurisdictions.

Transgender people are consistently discriminated against in jails and prisons. They are almost always put in the jail based on their sex and not their gender identity and are denied the ability to be able to take hormones or get sexual reassignment surgery and if they do they are never moved to prisons of the other gender. Studies show that trans people face an enormous amount of harassment and are at higher risk for sexual assault when they are not placed in prisons that match their gender identity.

Domestic Violence

Domestic violence takes a number of forms (including verbal, physical and psychological abuse), which vary across the gender spectrum. Domestic violence is tolerated and even legally accepted in many parts of the world. For instance, in 2010, the United Arab Emirates (UAE)'s Supreme Court ruled that a man has the right to physically discipline his wife and children if he does not leave physical marks.

According to a report of the Special Rapporteur submitted to the 58th session of the United Nations Commission on Human Rights (2002) concerning cultural practices in the family that reflect violence against women (E/CN.4/2002/83):

The Special Rapporteur indicated that there had been contradictory decisions with regard to the honour defence in Brazil, and that legislative provisions allowing for partial or complete defence in that context could be found in the penal codes of Argentina, Ecuador, Egypt, Guatemala, Iran, Israel, Jordan, Peru, Syria, Venezuela and the Palestinian National Authority.

Honour killings are a form of domestic violence which continues to be practiced in several parts of the world. The victims of honour killings are usually women. Victims of honour killings are killed for reasons such as refusing to enter an arranged marriage, being in a relationship that is disapproved by their relatives, having sex outside marriage, becoming the victim of rape, dressing in ways which are deemed inappropriate, or engaging in homosexual relations.

Stoning (a form of punishment in which a group throws stones at a person, usually until the victim dies) is associated with domestic disputes, especially those involving accusations of loss of chastity, adultery or the refusal of an arranged marriage. Recently, several people have been sentenced to death by stoning after being accused of adultery in Iran, Somalia, Afghanistan, Sudan, Mali and Pakistan.

Practices such as honour killings and stoning continue to be supported by mainstream politicians and other officials in some countries. In Pakistan, after the 2008 Balochistan honour killings in which five women were killed by tribesmen of the Umrani Tribe of Balochistan, Pakistani Federal Minister for Postal Services Israr Ullah Zehri defended the practice: "These are centuries-old traditions, and I will continue to defend them. Only those who indulge in immoral acts should be afraid." Following the 2006 case of Sakineh Mohammadi Ashtiani (which has placed Iran under international pressure for its stoning sentences), Mohammad-Javad Larijani (a senior envoy and chief of Iran's Human Rights Council) defended the practice of stoning; he claimed it was a "lesser punishment" than execution, because it allowed those convicted a chance at survival.

Dowry deaths are deaths of women or girls who are murdered or driven to suicide by continuous harassment and violence by husbands and in-laws in an effort to extort an increased dowry. Dowry deaths are most common in countries such as India, Pakistan, and Bangladesh. According to Amnesty International, "the ongoing reality of dowry-related violence is an example of what can happen when women are treated as property".

Education

The third Millennium Development Goal is directed at achieving gender equality and women's empowerment around the world. Improvement of equal educational opportunity contributes to the women's empowerment Women have traditionally had limited access to higher education. When women *were* admitted to higher education, they were encouraged to major in less-intellectual subjects; the study of English literature in American and British colleges and universities was instituted as a field considered suitable to women's "lesser intellects".

Educational specialities in higher education produce and perpetuate the existing segregation between men and women. Disparity persists particularly in computer and information science, where women receive only 21 percent of the undergraduate degrees, and in engineering, women obtain only 19 percent of the degrees in 2008.

World literacy is lower for females than for males. Latest data from CIA World Factbook shows that 79.7% of women are literate, compared to 88.6% of men (aged 15 and over). In some parts of the world, girls continue to be excluded from proper public or private education. In parts of Afghanistan, girls who go to school face serious violence from some local community members and religious groups. According to 2010 UN estimates, only Afghanistan, Pakistan and Yemen had less than 90 girls per 100 boys at school. Greater variation of gender educational discrepancy exists within countries. The situation appears to be more critical in rural areas, for instance UNICEF statistics revealed that in Nigerian villages less than 40 girls per 100 boys gained access to primary education. Studies of Sri Lanka economic development suggested that increased life expectancy of girls encourages educational investment because a longer time horizon increases the value of investments that pay out over time.

Girls' educational opportunities and results have greatly improved in the West. Since 1991, the proportion of women enrolled in college in the United States has exceeded the enrollment rate for men; the gap has widened over time. As of 2007, women made up the majority—54 percent—of the 10.8 million college students enrolled in the United States.

Research has found that discrimination continues; boys receive more attention, praise, blame and punishment in the grammar-school classroom, and "this pattern of more active teacher attention directed at male students continues at the postsecondary level". Over time, female students speak less in a classroom setting.

It is also argued that the educational system has become "feminized", allowing girls more of a chance at success with a more "girl-friendly" environment in the classroom; this is seen to hinder boys by punishing "masculine" behaviour and diagnosing boys with behavioural disorders.

In 2011 a United States Bureau of Labour Statistics report found that women are more likely than men to earn a bachelor's degree by age 23. A survey conducted by the Council of Graduate Schools similarly found that in the 2008–09 school year, women earned 50.9% of doctorates and 60% of master's degrees. However, only one-fifth of physics Ph.D.'s in the US are awarded to women, and only about half of those women are American; of all the physics professors in the country, only 14 percent are women.

There is a clear evidence that education plays an important role in determining women's input in financial and household decisions related to social and organisational matters.

Girls earn higher grades than boys until the end of high school; in some districts, girls achieve higher marks despite similar (or lower) scores than boys on standardised tests.

Fashion

Feminists argue that some fashion trends have been oppressive to women; they restrict women's movements, increase their vulnerability and endanger their health. The fashion industry is dealing with a great deal of criticism, as their association of thin-models and beauty has said to encourage bulimia and anorexia nervosa within women, as well as locking female consumers into false feminine identities.

The assigning of gender specific baby clothes from young ages is seen as sexist by some as it can instil in children from young ages a belief in strong gender stereotypes. An example of this is the assignment in some countries of the colour pink to girls and blue to boys. This fashion, however, is a recent one; at the beginning of the 20th century the trend was the opposite: blue for girls and pink for boys. In the early 1900s, The Women's Journal wrote: "That pink being a more decided and stronger colour, is more suitable for the boy, while blue, which is more delicate and dainty, is prettier for the girl." *DressMaker* magazine also explained: "The preferred colour to dress young boys in is pink. Blue is reserved for girls as it is considered paler, and the more dainty of the two colours, and pink is thought to be stronger (akin to red)."

Today, in most countries of the world, it is considered inappropriate for boys to wear dresses and skirts, but this, again, is a modern worldview: for example, from the mid-16th century until the late 19th or early 20th century, young boys in the Western world were unbreeched and wore gowns or dresses until an age that varied between two and eight. Also, throughout much of the history, in most cultures, men have worn skirt-like clothes and worn long hair. In many parts of the world, such as much of sub-Saharan Africa, it is considered normal for women to wear very short hair, with the hairstyles of boys being the same to those of girls.

Female Infanticide and Sex-Selective Abortion

Female infanticide is the killing of very young female children. It is an extreme form of gender based violence. Female infanticide is more common than male infanticide, and is especially prevalent in parts of Asia, such as parts of India and China. Recent studies suggest that over 90 million girls and women are missing in China and India as a result of systematic sex discrimination.

Sex-selective abortion involves terminating a pregnancy based upon the predicted gender of the baby. The abortion of female fetuses is most common in areas where the culture values male children over females, such as parts of the People's Republic of China, India, Pakistan, Korea, Taiwan, and the Caucasus. One reason for this preference is that males are seen as generating more income than females. A 2011 report on *Science Daily* stated that the trend grew steadily during the previous decade, and would probably cause a future shortage of women.

Forced Sterilization and Forced Abortion

Forced sterilization and forced abortion are forms of gender-based violence.

Forced sterilization was practiced during the first half of the 20th century by many Western countries (including the US) and there are reports of this practice being currently employed in some countries, such as Uzbekistan and China.

Forced abortions are today associated mostly with China. Although they are not an official policy of the country, they result from government pressure on local officials who, in turn, employ strong-arm tactics on pregnant mothers. In 2012, a highly publicized case of a forced abortion involving the photo of the fetus has sparked international outrage.

Female Genital Mutilation

Female genital mutilation (FGM), also known as female genital cutting and female circumcision, is defined by the World Health Organisation (WHO) as "all procedures that involve partial or total removal of the external female genitalia, or other injury to the female genital organs for non-medical reasons". WHO states that, "The procedure has no health benefits for girls and women" and "Procedures can cause severe bleeding and problems urinating, and later cysts, infections, infertility as well as complications in childbirth increased risk of newborn death" and "FGM is recognised internationally as a violation of the human rights of girls and women. It reflects deep-rooted inequality between the sexes, and constitutes an extreme form of discrimination against women". Feminist activist Seham Abd el Salam writes that FGM campaigners have been criticized by women's studies groups for their lack of recognition of male circumcision as a form of genital mutilation.

According to a 2013 UNICEF report, 125 million women and girls in Africa and the Middle East have experienced FGM. The highest prevalence in Africa is documented in Somalia (98 percent of women affected), Guinea (96 percent), Djibouti (93 percent), Egypt (91 percent), Eritrea (89 percent), Mali (89 percent), Sierra Leone (88 percent), Sudan (88 percent), Gambia (76 percent), Burkina Faso (76 percent), Ethiopia (74 percent), Mauritania (69 percent), Liberia (66 percent), and Guinea-Bissau (50 percent).

Hate-Motivated Sexual Assault

Rape and sexual assault are considered to be acts of hate. Their relationship to sexism is the frequent desire on the part of the perpetrator for power over the victim. The Centre for Women Policy Studies stated that "victims almost always are chosen for what they are rather than who they are"; a woman is more likely to be attacked because of her gender than her individuality.

Research into factors motivating perpetrators of rape against women frequently reveals a pattern of hatred towards women and pleasure in inflicting psychological and physical trauma, rather than sexual interest. Mary Odem posits that rape is the result not of pathology but of systems of male dominance, cultural practices and beliefs which objectify and degrade women.

Mary Odem, Jody Clay-Warner and Susan Brownmiller consider that sexist attitudes are propagated by a series of myths about rape and rapists. They state that in contrast to those myths, rapists often plan a rape before they choose a victim and acquaintance rape (not assault by a stranger) is the most common form of rape. Odem also asserts that these rape myths propagate sexist attitudes about men, by perpetuating the belief that men cannot control their sexuality.

In response to acquaintance rape, the "Men Can Stop Rape" movement has been implemented. The U.S. military has begun a similar movement, with the slogan "My strength is for defending".

In *Yes Means Yes: Visions of Female Sexual Power and A World Without Rape* by Jaclyn Friedman and Jessica Valenti, it is argued that in the US rape is most often popularly depicted (in the media, in public discourse, in movies etc.) as stranger-rape (with a stereotypical stranger who "jumps out of the bushes"), despite the fact that most rapes do not fit this stereotype. In presenting rape in this way to the public, patriarchal society ensures that women are kept under control, at home/indoors, living their life according to traditional gender roles and expectations, dressed 'modestly'. This leads to an ideology that women are safe at home (ignoring rape by family members/partners), and to blaming of victims.

Military Service

Military service has been considered a gender-specific duty. Some countries, such as Israel, require military service regardless of gender. Others (such as Finland, Turkey and Singapore) still use a system of conscription which only requires military service for men, although women are permitted to serve voluntarily.

In the United States, all men must register with the Selective Service System within 30 days of their 18th birthday. The system does not require women to register, leading to criticism that it discriminates against men by forcing them into a dangerous role based on gender.

Misandry

Misandry is the hatred of men and boys as a societal group or as individuals.

Misogyny

Misogyny is the hatred of women and girls as a societal group or as individuals.

Sexual Slavery

Sexual slavery occurs when people are coerced into slavery for sexual exploitation. The incidence of sexual slavery by country has been studied and tabulated by UNESCO (the United Nations Educational, Scientific and Cultural Organisation) with the cooperation of a number of international agencies. Sexual slavery also includes single-owner slavery, the ritual slavery associated with certain religious practices, slavery for primarily non-sexual purposes where non-consensual sex is common and forced prostitution.

Transphobia

Transphobia refers to prejudice against transsexuality and transsexual (or transgender) people based on their gender identification. Whether intentional or not, transphobia can have negative consequences for the object of the negative attitude. The lesbian, gay, bisexual and transgender (LGBT) movement opposes sexism against transsexuals. One form of sexism against transsexuals is "women-only" and "men-only" events and organisations, which have been criticized for excluding trans women and trans men.

Treatment of Victims of Rape

Stigmatization of women and girls who have been raped is widespread in many cultures; is a result of gender inequality and patriarchal social norms; and affects the ability of victims to recover.

In many parts of the world, women who have been raped are ostracized, rejected by their families, subjected to violence - in extreme cases to honour killings, because they are deemed to have brought 'shame' on their families.

There is also a strong connection between rape and forced marriage, through practices such as forcing of a woman or girl who has been raped to marry her rapist, in order to restore the 'honour' of her family; or marriage by abduction, a practice in which a man abducts the woman or girl whom he wishes to marry and rapes her, in order to force the marriage (this practice is very common in Ethiopia). The criminalization of marital rape (forcing one's own spouse to have sex) is very recent, having occurred during the past few decades; and in many countries it is still legal. Several countries in Eastern Europe and Scandinavia made spousal rape illegal before 1970; other countries in Western Europe and the English-speaking Western World outlawed it later, mostly in the 1980s and 1990s. In some parts of the world the lack of criminalization of marital rape coupled with the practice of child marriage leads to serious forms of child sexual abuse being legitimized. The WHO wrote that: "Marriage is often used to legitimize a range of forms of sexual violence against women. The custom of marrying off young children, particularly girls, is found in many parts of the world. This practice – legal in many countries – is a form of sexual violence, since the children involved are unable to give or withhold their consent".

In countries where fornication or adultery are illegal, victims of rape can be charged under this laws (even if the victims succeed in proving their rape case, they can still be charged with a criminal offence if the court finds they were not virgins at the time of the assault - if they were unmarried).

The WHO states that: "Persons who have been raped should receive treatment consistent with the dignity and respect they are owed as human beings."

War Rape

War rapes are rapes committed by soldiers, other combatants or civilians during armed conflict, war or military occupation, and are distinguished from in-service sexual assault and rape committed amongst troops. It also encompasses situations in which women are forced into prostitution or sexual slavery by an occupying power, such as Japanese comfort women during World War II. Sexual violence and rape as a weapon of war also affects a large number of men, although it is less-often reported. The war-torn Democratic Republic of the Congo has been described as the "rape capital of the world" with more than 400,000 rapes reported in just one year.

Antisexism Movements

In Ecuador, the Pink Helmets (*Cascos Rosas*), created in May 2011 by 18-year-olds Freddy Caleron and Damian Valencia, sought to unite young men against machismo. The Pink Helmets published a manual with suggestions to end violence in society, relationships, within families, and among friends. The Pink Helmets' founders and participating members, who call themselves *neomasculinos* and are 15–20 years old, visit schools and the streets to give talks and participate in festivals (such as the Quito Fest) in collaboration with UN Women (part of the United Nations) to increase awareness of violence and the movement against it. Chauvinism and violence are common in Ecuador, with 40% of children under age 15 saying that they have witnessed acts of violence at home. According to the *National Plan for the Eradication of Gender Violence of the Government of Ecuador*, 80% of women have been victims of violence at least once and 21% of children and adolescents have been sexually abused (no official data provide a concrete number of women who have been killed or injured by men). The movement aims to promote gender equality, eliminate gender violence against women and encourage a "dialogue of respect between men and women from adolescence without hesitation to criticize sexist concepts that live in their culture".

Ageism

Ageism (also spelled "agism") is stereotyping and discriminating against individuals or groups on the basis of their age. This may be casual or systematic. The term was coined in 1971 by Robert Neil Butler to describe discrimination against seniors, and patterned on sexism and racism. Butler defined "ageism" as a combination of three connected elements. Among them were prejudicial attitudes towards older people, old age, and the aging process; discriminatory practices against older people; and institutional practices and policies that perpetuate stereotypes about older people.

The term has also been used to describe prejudice and discrimination against adolescents and children, including ignoring their ideas because they are too young, or assuming that they should behave in certain ways because of their age.

Ageism in common parlance and age studies usually refers to negative discriminatory practices against old people, people in their middle years, teenagers and children. There are several forms of age-related bias. Adultism is a predisposition towards adults, which is seen as biased against children, youth, and all young people who are not addressed or viewed as adults. Jeunism is the discrimination against older people in favour of younger ones. This includes political candidacies, jobs, and cultural settings where the supposed greater vitality and/or physical beauty of youth is more appreciated than the supposed greater moral and/or intellectual rigor of adulthood. Adultcentricism is the "exaggerated egocentrism of adults." *Adultocracy* is the social convention which defines "maturity" and "immaturity," placing adults in a dominant position over young people, both theoretically and practically. Gerontocracy is a form of oligarchical rule in which an entity is ruled by leaders who are significantly older than most of the adult population. Chronocentrism is primarily the belief that a certain state of humanity is superior to all previous and/or future times.

Based on a conceptual analysis of ageism, a new definition of ageism was introduced by Iversen, Larsen, & Solem in 2009. This definition constitutes the foundation for higher reliability and validity in future research about ageism and its complexity offers a new way of systemizing theories on ageism: "Ageism is defined as negative or positive stereotypes, prejudice and/or discrimination against (or to the advantage of) elderly people on the basis of their chronological age or on the basis of a perception of them as being 'old' or 'elderly'. Ageism can be implicit or explicit and can be expressed on a micro-, meso- or macro-level" (Iversen, Larsen & Solem, 2009).

Other conditions of fear or aversion associated with age groups have their own names, particularly: paedophobia, the fear of infants and children; ephebiphobia, the fear of youth, sometimes also referred to as an irrational fear of adolescents or a prejudice against teenagers; and gerontophobia, the fear of elderly people.

Implicit Ageism

Implicit ageism is the term used to refer to the implicit or subconscious thoughts, feelings, and behaviours one has about older or younger people. These may be a mixture of positive and negative thoughts and feelings, but gerontologist Becca Levy reports that they "tend to be mostly negative."

Ageist Stereotyping

Ageist stereotyping is a tool of cognition which involves categorizing into groups and attributing characteristics to these groups. Stereotypes are necessary for processing huge volumes of information which would otherwise overload a person, and they are often based on a "grain of truth" (for example, the association between aging and ill health). However, they cause harm when the content of the stereotype is incorrect with respect to most of the group or where a stereotype is so strongly held that it overrides evidence which shows that an individual does not conform to it. For example, age-based stereotypes prime one to draw very different conclusions when one sees an older and a younger adult with, say, back pain or a limp. One might well assume that the younger person's condition is temporary and treatable, following an accident, while the older person's condition is chronic and less susceptible to intervention. On average, this might be true, but plenty of older people have accidents and recover quickly and very young people (such as infants, toddlers and small children) can become permanently disabled in the same situation. This assumption may have no consequence if one makes it in the blink of an eye as one is passing someone in the street, but if it is held by a health professional offering treatment or managers thinking about occupational health, it could inappropriately influence their actions and lead to age-related discrimination. Managers have been accused, by Erdman Palmore, as stereotyping older workers as being resistant to change, not creative, cautious, slow to make judgments, lower in physical capacity, uninterested in technological change, and difficult to train. Another example is when people are rude to children because of their high pitched voice, even if they are kind and courteous. A review of the research literature related to age stereotypes in the workplace was recently published in the Journal of Management.

Ageist Prejudice

Ageist prejudice is a type of emotion which is often linked to the cognitive process of stereotyping. It can involve the expression of derogatory attitudes, which may then lead to the use of discriminatory behaviour. Where older or younger contestants were rejected in the belief that they were poor performers, this could well be the result of stereotyping. But older people were also voted for at the stage in the game where it made sense to target the best performers. This can only be explained by a subconscious emotional reaction to older people; in this case, the prejudice took the form of distaste and a desire to exclude oneself from the company of older people.

Benevolent Prejudice

Stereotyping and prejudice against different groups in society does not take the same form. Age-based prejudice and stereotyping usually involves older or younger people being pitied, marginalized, or patronized. This is described as "benevolent prejudice" because the tendency to pity is linked to seeing older or younger people as "friendly" but "incompetent." This is similar to the prejudice most often directed against women and disabled people. Age Concern's survey revealed strong evidence of "benevolent prejudice." 48% said that over-70s are viewed as friendly (compared to 27% who said the same about under-30s). Meanwhile, only 26% believe over-70s are viewed as capable (with 41% saying the same about under-30s).

The figure for friendliness of under-30s is, conversely, an example of Hostile Prejudice.

Hostile Prejudice

"Hostile prejudice" based on hatred, fear, aversion, or threat often characterizes attitudes linked to race, religion, disability, and sex. An example of hostile prejudice toward youth is the presumption without any evidence that a given crime was committed by a young person. Rhetoric regarding intergenerational competition can be motivated by politics. Violence against vulnerable older people can be motivated by subconscious hostility or fear; within families, this involves impatience and lack of understanding. Equality campaigners are often wary of drawing comparisons between different forms of inequality. But it is unquestionably true that abuse and neglect experienced by vulnerable older people (which is closely linked to hostile prejudice) kills more people each year than the shocking but relatively isolated cases of public violence motivated by race, religion, or sexual orientation.

The impact of "benevolent" and "hostile" prejudice tends to be different. The warmth felt towards older or younger people and the knowledge that many have no access to paid employment means there is often public acceptance that they are deserving of preferential treatment—for example, less expensive movie and bus fares. But the perception of incompetence means older and younger people can be seen as "not up to the job" or "a menace on the roads," when there is little or exaggerated evidence to support this. Prejudice also leads to assumptions that it is "natural" for older or younger people to have lower expectations, reduced choice and control, and less account taken of their views.

Discrimination

Age discrimination refers to the actions taken to deny or limit opportunities to people on the basis of age. These are usually actions taken as a result of one's ageist beliefs and attitudes. Age discrimination occurs on both a personal and institutional level.

On a personal level, an older person may be told that he or she is too old to engage in certain physical activities, like an informal game of basketball between friends and family. A younger person may be told they are too young to get a job or help move the dining room table. On an institutional level, there are policies and regulations in place that limit opportunities to people of certain ages and deny them to all others. The law, for instance, requires that all young persons must be at least 16 years old in order to obtain a driver's license in the United States. There are also government regulations that determine when a worker may retire. Currently, in the US, a worker must be between 65 and 67 years old (depending upon his or her birth year) before becoming eligible for Social Security retirement benefits, but some company pension plans begin benefits at earlier ages.

A 2006/2007 survey done by the Children's Rights Alliance for England and the National Children's Bureau asked 4,060 children and young people whether they have ever been treated unfairly based on various criteria (race, age, sex, sexual orientation, etc.). A total of 43% of British youth surveyed reported experiencing discrimination based on their age, far eclipsing other categories of discrimination like sex (27%), race (11%), or sexual orientation (6%). Consistently, a study based on the European Social Survey found that whereas 35% of Europeans reported exposure ageism, only 25% reported exposure to sexism and as few as 17% reported exposure to racism.

Ageism has significant effects in two particular sectors: employment and health care.

Employment

The concept of ageism was originally developed to refer to prejudice and discrimination against older people and middle age, but has expanded to include children and teenagers.

Like racial and gender discrimination, age discrimination, at least when it affects younger workers, can result in unequal pay for equal work. Unlike racial and gender discrimination, however, age discrimination in wages is often enshrined in law. For example, in both the United States and the United Kingdom minimum wage laws allow for employers to pay lower wages to young workers. Many state and local minimum wage laws mirror such an age-based, tiered minimum wage. Midlife workers, on average, make more than younger workers do, which reflects educational achievement and experience of various kinds (job-specific, industry-specific, etc.). The age-wage peak in the United States, according to Census data, is between 45 and 54 years of age. Seniority in general accords with respect as people age, lessening ageism.

Statistical discrimination refers to limiting the employment opportunities of an individual based on stereotypes of a group to which the person belongs. Limited employment opportunities could come in the form of lower pay for equal work or jobs with little social mobility. Younger female workers were historically discriminated against, in comparison with younger men, because it was expected that, as young women of childbearing years, they would need to leave the work force permanently or periodically to have children.

Labour regulations also limit the age at which people are allowed to work and how many hours and under what conditions they may work. In the United States, a person must generally be at least 14 years old to seek a job, and workers face additional restrictions on their work activities until they reach age 16. Many companies refuse to hire workers younger than 18.

While older workers benefit more often from higher wages than do younger workers, they face barriers in promotions and hiring. Employers also encourage early retirement or layoffs disproportionately more for older or more experienced workers.

Age discrimination in hiring has been shown to exist in the United States. The Equal Employment Opportunity Commission's first complainants were female flight attendants complaining of (among other things) age discrimination. More recently, Joanna Lahey, professor at The Bush School of Government and Public Service at Texas A&M, found that firms are more than 40% more likely to interview a young adult job applicant than an older job applicant.

In a survey for the University of Kent, England, 29% of respondents stated that they had suffered from age discrimination. This is a higher proportion than for gender or racial discrimination. Dominic Abrams, social psychology professor at the university, concluded that Ageism is the most pervasive form of prejudice experienced in the UK population.

According to Dr. Robert M. McCann, an associate professor of management communication at the University of Southern California's Marshall School of Business, denigrating older workers, even if only subtly, can have an outsized negative impact on employee productivity and corporate profits. For American corporations, age discrimination can lead to significant expenses. In Fiscal Year 2006, the U.S. Equal Employment Opportunity Commission received nearly 17,000 charges of age discrimination, resolving more than 14,000 and recovering $51.5 million in monetary benefits. Costs from lawsuit settlements and judgments can run into the millions, most notably with the $250 million paid by the California Public Employees' Retirement System (CalPERS) under a settlement agreement in 2003.

In the UK, age discrimination against older people has been prohibited in employment since 2006. Since then, the number of age discrimination cases rose dramatically. The laws protect the anyone over the age of 16 who is young as well as old. There were over 6,800 claims submitted to the Employment Tribunal in 2010/11 compared with just 900 in 2006/2007 (immediately after the Regulations came in force). However, the figures for 2011/2012 show a 47% fall in the number of claims, and commentators have suggested that the repeal of the Default Retirement Age may be the reason behind this.

Some political offices have qualifications that discriminate on the basis of age as a proxy for experience, education, or accumulated wisdom. For example, the President of the United States must be at least 35 years old; a United States Senator must be at least 30; and a United States Congressman must be at least 25.

Healthcare

There is considerable evidence of discrimination against the elderly in health care. This is particularly true for aspects of the physician-patient interaction, such as screening procedures, information exchanges, and treatment decisions. In the patient-physician interaction, physicians and other health care providers may hold attitudes, beliefs, and behaviours that are associated with Ageism against older patients. Studies have found that some physicians do not seem to show any care or concern toward treating the medical problems of older people. Then, when actually interacting with these older patients on the job, the doctors sometimes view them with disgust and describe them in negative ways, such as "depressing" or "crazy." For screening procedures, elderly people are less likely than younger people to be screened for cancers and, due to the lack of this preventative measure, less likely to be diagnosed at early stages of their conditions.

After being diagnosed with a disease that may be potentially curable, older people are further discriminated against. Though there may be surgeries or operations with high survival rates that might cure their condition, older patients are less likely than younger patients to receive all the necessary treatments. It has been posited that this is because doctors fear their older patients are not physically strong enough to tolerate the curative treatments and are more likely to have complications during surgery that may end in death. However, other studies have been done with patients with heart disease, and, in these cases, the older patients were still less likely to receive further tests or treatments, independent of the severity of their health problems. Thus, the approach to the treatment of older people is concentrated on managing the disease rather than preventing or curing it. This is based on the stereotype that it is the natural process of aging for the quality of health to decrease, and, therefore, there is no point in attempting to prevent the inevitable decline of old age.

Differential medical treatment of elderly people can have significant effects on their health outcomes, a differential outcome which somehow escapes established protections against Ageism.

Effects of Ageism

Ageism has significant effects on the elderly and young people. The stereotypes and infantilization of older and younger people by patronizing language affects older and younger people's self-esteem and behaviours. After repeatedly hearing a stereotype that older or younger people are useless, older and younger people may begin to feel like dependent, non-contributing members of society. They may start to perceive themselves in terms of the looking-glass self—that is, in the same ways that others in society see them. Studies have also specifically shown that when older and younger people hear these stereotypes about their supposed incompetence and uselessness, they perform worse on measures of competence and memory. These stereotypes then become self-fulfilling prophecies. According to Becca Levy's Stereotype Embodiment Theory, older and younger people might also engage in self-stereotypes, taking their culture's age stereotypes—to which they have been exposed over the life course—and directing them inward toward themselves. Then this behaviour reinforces the present stereotypes and treatment of the elderly.

Many overcome these stereotypes and live the way they want, but it can be difficult to avoid deeply ingrained prejudice, especially if one has been exposed to ageist views in childhood or adolescence.

Australia

Australia has had age discrimination laws for some time. Discrimination on the basis of age is illegal in each of the states and territories of Australia. At the national level, Australia is party to a number of international treaties and conventions that impose obligations to eliminate age discrimination.

The Australian Human Rights Commission Act 1986 established the Australian Human Rights Commission and bestows on this Commission functions in relation to a number of international treaties and conventions that cover age discrimination. During 1998-1999, 15% of complaints received by the Commission under the Act were about discrimination on the basis of age.

Age discrimination laws at the national level were strengthened by the Age Discrimination Act 2004, which helps to ensure that people are not subjected to age discrimination in various areas of public life, including employment, the provision of goods and services, education, and the administration of Australian government laws and programs. The Act, however, does provide for exemptions in some areas, as well as providing for positive discrimination, that is, actions which assist people of a particular age who experience a disadvantage because of their age.

In 2011, for the first time a position of Age Discrimination Commissioner was created within the Australian Human Rights Commission. The new Commissioner's responsibilities include raising awareness among employers about the beneficial contributions that senior Australians as well as younger employees can make in the workforce.

Canada

In Canada, Article 718.2, clause (a)(i), of the Criminal Code defines as aggravating circumstances, among other situations, "evidence that the offence was motivated by ... age".

Mandatory retirement was ended in Canada in December 2011, but 74% of Canadians still consider age discrimination to be a problem.

Nigeria

In November 2011, the Nigerian House of Representatives considered a bill which would outlaw age discrimination in employment.

United States

In the US, each state has its own laws regarding age discrimination, and there are also federal laws. In California, the Fair Employment and Housing Act forbids unlawful discrimination against persons age 40 and older. The FEHA is the principal California statute prohibiting employment discrimination, covering employers, labour organisations, employment agencies, apprenticeship programs and/or any person or entity who aids, abets, incites, compels, or coerces the doing of a discriminatory act. In addition to age, it prohibits employment discrimination based on race or colour; religion; national origin or ancestry, physical disability; mental disability or medical condition; marital status; sex or sexual orientation; and pregnancy, childbirth, or related medical conditions. Although there are many protections for age-based discrimination against older workers (as shown above) there are very few similar protections for younger workers.

The District of Columbia and twelve states define age as a specific motivation for hate crimes – California, Florida, Iowa, Hawaii, Kansas, Louisiana, Maine, Minnesota, Nebraska, New Mexico, New York and Vermont.

The federal government governs age discrimination under the Age Discrimination in Employment Act of 1967 (ADEA). The ADEA prohibits employment discrimination based on age with respect to employees 40 years of age or older as well. The ADEA also addresses the difficulty older workers face in obtaining new employment after being displaced from their jobs, arbitrary age limits. The ADEA applies even if some of the minimum 20 employees are overseas and working for a US corporation.

The United States federal government has responded to issues of youth-bias in governance through several measures in the past. They include the creation of the 1970s-era National Commission on Resources for Youth, which was created in the late 1960s as to promote youth participation throughout communities. Recently the federal government implemented the Tom Osborne Federal Youth Coordination Act, aiming to curb redundancy among federal service providers to youth.

The Children's Online Privacy Protection Act has been criticized by some as ageist.

As it approaches its fiftieth anniversary, one author argues that America's leading law against age discrimination, the Age Discrimination in Employment Act of 1967, offers little real protection to older workers. In her 2014 book, Patricia G. Barnes, an attorney and judge, argues the ADEA was riddled with loopholes to begin with and has been eviscerated over time by the U.S. Supreme Court. Moreover, her book, *Betrayed: The Legalization of Age Discrimination in the Workplace*, argues the "legalization" of age discrimination has caused a trickle down affect. Workers who are otherwise considered to be young - workers in their 30s, 40s and 50s - are increasingly experiencing age discrimination.

European Union

The European citizenship provides the right to protection from discrimination on the grounds of age. According to Article 21-1 of the Charter of Fundamental Rights of the European Union s:Charter of Fundamental Rights of the European Union.

Additional protection against age discrimination comes from the Framework Directive 2000/78/EC. It prohibits discrimination on grounds of age in the field of employment.

Germany

On 18 August 2006, the General Equal Treatment Act (Allgemeines Gleichbehandlungsgesetz, AGG) came into force. The aim of the AGG is to prevent and abolish discrimination on various grounds including age. A recent study suggested that youths in Germany feel the brunt of age discrimination.

France

In France, Articles 225-1 through 225-4 of the Penal Code detail the penalization of Ageism, when it comes to an age discrimination related to the consumption of a good or service, to the exercise of an economic activity, to the labour market or an internship, except in the cases foreseen in Article 225-3.

Belgium

In Belgium, the Law of 25 Feb 2003 "tending to fight discrimination" punishes Ageism when "a difference of treatment that lacks objective and reasonable justification is directly based on ... age". Discrimination is forbidden when it refers to providing or offering a good or service, to conditions linked to work or employment, to the appointment or promotion of an employee, and yet to the access or participation in "an economic, social, cultural or political activity accessible to the public" (Article 2nd, § 4). Incitement to discrimination, to hatred or to violence against a person or a group on the grounds of (...) age (Article 6) is punished with imprisonment and/or a fine. Nevertheless, employment opportunities are worsening for people in their middle years in many of these same countries, according to Martin Kohli *et al.* in *Time for Retirement* (1991).

United Kingdom

In the UK, laws against Ageism are new. Age discrimination laws were brought into force in October 2006, and can now be found in the Equality Act 2010. This implements the Equal Treatment Framework Directive 2000/78/EC and protects employees against direct discrimination, indirect discrimination, harassment and victimisation. There is also provision in the Equality Act 2010 to prohibit age discrimination in the provision of goods and services, though this has not yet been implemented by the current UK Coalition Government and will not be implemented before October 2012 at the earliest.

Despite the relatively recent prohibition on age discrimination, there have already been many notable cases and official statistics show a 37% increase in claims in 2009/10 and a further 31% increase in 2010/11. Examples include the case involving Rolls Royce, the "Heyday" case brought by Age UK and the recent Miriam O'Reilly case against the BBC.

Recent research suggested that the number of age discrimination claims annually could reach 15,000 by 2015.

The European Social Study survey in 2011 revealed that nearly two out of five people claim to have been shown a lack of respect because of their age. The survey suggested that the UK is riven by intergenerational splits, with half of people admitting they do not have a single friend over 70; this compares with only a third of Portuguese, Swiss and Germans who say that they do not have a friend of that age or older. A Demos study in 2012 showed that three quarters of people in the UK believed there to be not enough opportunities for older and younger people to meet and work together.

The "Grey Pride" campaign has been advocating for a Minister for Older People and its campaign has had some success, with Labour Leader Ed Miliband appointing Liz Kendall as Shadow Minister for Older People. The artist Michael Freedman, an outspoken advocate against age discrimination within the art world says that "mature students, like me, come to art late in life, so why are we penalised and demotivated? Whatever happened to lifelong learning and the notion of a flexible workforce?"

Advocacy Campaigns

Many current and historical intergenerational and youth programs have been created to address the issue of Ageism. Among the advocacy organisations created in the United Kingdom to challenge age discrimination are Age UK and the British Youth Council.

In the United States there have been several historic and current efforts to challenge Ageism. The earliest example may be the Newsboys Strike of 1899, which fought ageist employment practices targeted against youth by large newspaper syndicates in the Northeast. During the Franklin D. Roosevelt Administration, First Lady Eleanor Roosevelt was active in the national youth movement, including the formation of the National Youth Administration and the defence of the American Youth Congress. She made several statements on behalf of youth and against Ageism. In one report entitled, "Facing the Problems of Youth," Roosevelt said of youth,

> *"We cannot simply expect them to say, 'Our older people have had experience and they have proved to themselves certain things, therefore they are right.' That isn't the way the best kind of young people think. They want to experience for themselves. I find they are perfectly willing to talk to older people, but they don't want to talk to older people who are shocked by their ideas, nor do they want to talk to older people who are not realistic."*

Students for a Democratic Society formed in 1960 to promote democratic opportunities for all people regardless of age, and the Gray Panthers was formed in the early 1970s with a goal of eliminating Ageism in all forms. Three O'Clock Lobby formed in 1976 to promote youth participation throughout traditionally ageist government structures in Michigan, while Youth Liberation of Ann Arbor started in 1970 to promote youth and fight Ageism.

More recent U.S. programs include Americans for a Society Free from Age Restrictions, which formed in 1996 to advance the civil and human rights of young people through eliminating ageist laws targeted against young people, and to help youth counter Ageism in America. The National Youth Rights Association started in 1998 to promote awareness of the legal and human rights of young people in the United States, and the Freechild Project was formed in 2001 to identify, unify and promote diverse opportunities for youth engagement in social change by fighting Ageism.

Related Campaigns

- In 2002 the Writers Guild of America, West has waged a legal battle within the entertainment industry to eliminate age discrimination commonly faced by elder scriptwriters.
- Director Paul Weitz reported he wrote the 2004 film, *In Good Company* to reveal how Ageism affects youth and adults.
- In 2002 The Freechild Project created an information and training initiative to provide resources to youth organisations and schools focused on youth rights.
- In 2006 Lydia Giménez-LLort, PhD an assistant professor of Psychiatry and researcher at the Autonomous University of Barcelona coined the term 'Snow White Syndrome' at the 'Congrés de la Gent Gran de Cerdanyola del Vallès' (Congress of the Elderly of Cerdanyola del Vallès, Barcelona, Spain) as a metaphor to define Ageism in an easier and more friendly way while developing a constructive spirit against it. The metaphor is based on both the auto-Ageism and adultocracy exhibited by the queen of the Snow White fairy tale as well as the social Ageism symbolized by the mirror
- Since 2008 'The Intergenerational Study' by Lydia Giménez-LLort and Paula Ramírez-Boix from the Autonomous University of Barcelona is aimed to find the basis of the link between grandparents and grandsons (positive family relationships) that are able to minimize the Ageism towards the elderly. Students

of several Spanish universities have enrolled in this study which soon will be also performed in USA, Nigeria, Barbados, Argentina and Mexico. The preliminary results reveal that 'The Intergenerational study questionnaire' induces young people to do a reflexive and autocritic analysis of their intergenerational relationships in contrast to those shown towards other unrelated old people which results very positive to challenge Ageism. A cortometrage about 'The International Study' has been directed and produced by Tomás Sunyer from Los Angeles City College

- Votes at 16 intends to lower the voting age to 16, reducing Ageism and giving 16 year olds equal pay on the National Minimum Wage. The group claims that 16 year olds get less money than older people for the same work, angering many 16 year olds. They additionally postulate that 16 year olds will have their voice listened to by older people more often.

Accusations of Ageism

In a recent interview, actor Pierce Brosnan cited ageism as one of the contributing factors as to why he was not asked to continue his role as James Bond in the Bond film *Casino Royale*, released in 2006.

Also, successful singer and actress, Madonna spoke out in her 50s about ageism and her fight to defy the norms of society. Similarly, Sex and the City star Kim Cattrall has also raised the issue of ageism.

A 2007 Pew Research Centre study found that a majority of American voters would be less likely to vote for a President past a given age, with only 45% saying that age would not matter.

Economy and Society

Economic Development: Economic development is the development of economic wealth of countries or regions for the well-being of their inhabitants. Public policy generally aims at continuous and sustained economic growth and expansion of national economies so that developing countries become developed countries. The economic development process supposes that legal and institutional adjustments are made to give incentives for innovation and for investments so as to develop an efficient production and distribution system for goods and service.

Economic System of Simple Societies

Herbert Spencer has defined simple society as one which forms a simple working whole unsubjected to any other and of which the parts cooperate for certain public ends. Simple societies have low division of labour. The occupational differentiation being limited primarily to birth, sex and age. These societies have no specialised economic organisation.

The productive skills are simple and productivity is low therefore these societies cannot sustain large population size. Most of the adult members are engaged in food gathering activities. There is little or no surplus so the social inequalities are not significant and economic interaction takes place within egalitarian framework.

The production system is simple but exchange of goods and services assume a complex form. The forms of exchange are reciprocal and redistributive type. Some of the simple societies inhabiting regions having abundant food and other resources indulge in conspicuous consumption. The members lack high degree of achievement motivation as there is neither any intense preoccupation on generation and accumulation of economic surplus. Infact most economic activities emphasize on giving rather than storing or accumulation. Private ownership of means of production is non-existent.

There is no clear separation between domestic economy and community economy as they overlap to varying degrees.

The economic system is dominated by sacred consisting of magic-religious ideas.

The innovation is rare and change is slow. The customary practices and norms regulate production and exchange of goods and services.

Some Forms of Simple Economic Exchange

Barter System: It is direct form of exchange whether in return for services or goods.

Silent Trade: It was an exchange system where the exchanging parties do not know each other personally.

Jajmani System: It is system of economic and social relationship existing between various castes in villages. The patron is known as jajman and the service castes are known as kamin. It is still prevalent in villages.

Ceremonial Exchange: It is a type of social system in which goods are given to relatives and friends on various social occasions. The main idea is to establish cordial relations between the various social groups.

Potlatch: This term means gift. It is meant as a public distribution of goods made to establish certain claims of the giver and the recipients. It is based on the principle of reciprocity. Through this system the host declares his status to others.

Multicentric Economy: It is an economy using several media of exchange.

Kula: According to Malinowski it is a ceremonial exchange participated by the inhabitants of a closed circle of Trobriand Island. It has no practical or commercial value. The system of exchange is regulated in a kind of ring with two directional movements. In clockwise direction, the red shell necklaces called Soulava circulate and in anticlockwise circulation the white arm shells known as Mwali circulate among the members of the Kula. Objects given and taken in Kula are never subjected any bargaining.

Economic System of Complex Societies

The complex societies have high degree of division of labour and consequently structural differentiation. Thus economic activity constitutes a specialised activity taking place in special institution framework and distinguishable from other types of social activity e.g. factories, banks and markets are some of the distinct economic activities. High division of labour implies advanced skills which help in high productivity. The economic organisation can easily sustain a large population. Complex societies due to their high productivity generate huge surplus. They can support conspicuous consumption.

Market exchange is the pivotal form of exchange and money is the universal medium of exchange. The members of the complex societies have high achievement motivation and the economic behaviour is characterized by an intense preoccupation with generation and accumulation of surplus. There exist a clear distinction between domestic economy and community economy. The domestic units are the units of consumption and supply the manpower to the community economy. The production of goods and services takes place in the larger units which form part of the community economy.

These societies are characterized by the high level of scientific and technological advancements. Economic activity is perceived in secular terms and is based on practical rationality. High degree of specialisation, rapidity of change, predominance of practical and excessive mechanization of production leads to a state of anomie in society and alienate the worker from the product of his labour.

Market Economy

Market or Free economy is characterized by a system in which the allocation of resources is determined by supply and demand in the market. Both the production and distribution is determined by the market forces to ensure competition and efficiency .It has an effect on the traditional families as a result of monetisation and market economy the different members of the family contribute to the family income and increased the avenues for social mobility. Their is rapid growth of industries in which the employee-employer relations are based on contractual relations. Work has become the commodity which is exchanged for wages. Expansion of markets has increased the volume of trade and commerce facilitating the integration of the country. Growth of economy leads to occupational diversification and increasing specialisation of occupations which in turn has created a demand for educational institutions to provide specialised training. Due to industrialization and expansion of market economy in urban areas leads to consumption oriented life-style. Market economy governed by supply and demand is inherently unstable. This leads to anomie which is characteristic of urban life. Inflation also poses constant threat to instability in the urban markets.

Planned Economy

A planned economy is an economic system in which decisions about the production, allocation and consumption of goods and services is planned ahead of time, in either a centralized or decentralized fashion. Since most known planned economies rely on plans implemented by the way of command, they have become widely known as command economies. The government takes the initiative and set the goals and targets to be followed by the market forces. The state intervention is limited to formulation of plan and adoption of indirect controls. The private sector becomes partner in the formulation of a plan and responsible for its implementation.

Social Determinants of Economic Development

Economic development implies two things: Economic growth which leads to increase in production and generation of income and equitable distribution of this income among the population to improve the quality of life. Although economic development does not necessarily imply industrialization there is no historical precedent for substantial increase in percapita income without diversion of both capital and labour from agriculture. Economic development is synonymous with industrialization. Economic development is very much influenced by various social factors. Nation states are created with common language and culture. Economic development of any country hinges on the efficient employment of factors of production such as labour, land, capital and organisation. There is commercialization of production with monetization of economy. The employment of factors of production is conditioned by cultural and social factors. The people must have the required ability, experience and knowledge to make the best use of the facilities that are made available. There is decline of the proportion of the working population engaged in agriculture. The technology plays very important role when appropriate social conditions are present.

There is trend towards urbanization of society with growth of scientific knowledge. A new value system emerges which emphasis individual initiative and responsibility and enables the individual to function without any control. The exclusiveness of clan, kin or caste breaks down and provides norms of behaviour suited to the secondary group type of relationship characteristic of an industrial society. There is widespread spread of education. The social stratification emerges based on achievement criteria and permitting occupational mobility.

Concept of Property

Property refers to the rights that the owner of the object has in relation to others who are not owners of the object. Property rights are backed by the state and enforced through its legal institutions. According to Morris Ginsberg property may be described as the set of rights and obligations which define the relations between individuals or groups in respect of their control over material things or persons treated as things. Kingsley Davis defines property as consisting of rights and duties of one person or group as against all other persons and groups with respect to some scarce good. It is thus exclusive for it sets off what is mine from what is thine but it is also social being rooted in custom and protected by law.

Some Theoretical Concepts

Karl Marx: Karl Marx has distinguished between different types of societies on basis of economic system. These are primitive communism, ancient slave production, feudalism and capitalism, socialism and communism. A man is both the producer and product of society. Marx's analysis of history is based on his distinction between the means/forces of production literally those things, such as land, natural resources, and technology, that are necessary for the production of material goods, and the relations of production in other words, the social and technical relationships people enter into as they acquire and use the means of production. Together these comprise the mode of production. Marx observed that within any given society the mode of production changes, and that European societies had progressed from a feudal mode of production to a capitalist mode of production. Marx did not understand classes as purely subjective. He sought to define classes in terms of objective criteria, such as their access to resources. For Marx, different classes have divergent interests, which is another source of social disruption and conflict. Marx was especially concerned with how people relate to that most fundamental resource of all, their own labour power. Marx wrote extensively about this in terms of the problem of alienation. For Marx, the possibility that one may give up ownership of one's own labour - one's capacity to transform the world - is tantamount to being alienated from one's own nature; it is a spiritual loss.

Marx described this loss in terms of commodity fetchism, in which the things that people produce, commodities, appear to have a life and movement of their own to which humans and their behaviour merely adapt. This disguises the fact that the exchange and circulation of commodities really are the product and reflection of social relationships among people. Under capitalism, social relationships of production, such as among workers or between workers and capitalists, are mediated through commodities, including labour, that are bought and sold on the market. According to Marx, a capitalist mode of production developed in Europe when labour itself became a commodity - when peasants became free to sell their own labour-power, and needed to do so because they no longer possessed their own land or tools necessary to produce. People sell their labour-power when they accept compensation in return for whatever work they do in a given period of time (in other words, they are not selling the product of their labour, but their capacity to work). In return for selling their labour power they receive money, which allows them to survive. Those who must sell their labour power to live are "proletariat. The person who buys the labour power, generally someone who does own the land and technology to produce, is a "capitalist" or "bourgeoise. The capitalist mode of production is capable of tremendous growth because the capitalist can, and has an incentive to, reinvest profits in new technologies. Marx considered the capitalist class to be the most revolutionary in history, because it constantly revolutionized the means of production.

Max Weber

Max Weber formulated a three component theory of social stratification with social class, status class and party class (or political class) as conceptually distinct elements. Social class is based on economically determined relationship to the market (owner, employee etc.). Status class is based on non-economical qualities like honour, prestige and religion. Party class refers to affiliations in the political domain. All three dimensions have consequences for what Weber called "life chances". According to Weber there are two sources of power. One is derived from constellation of interests that develop in a free market and the other is from an established system of authority that allocates the right to command and the duty to obey.

Emile Durkheim

Emile Durkheim sees division of labour in terms of social process. He has tried to determine the social consequences of the division of labour in the modern societies. He has made a fundamental difference between pre-industrial and industrial societies and also made difference between two types of solidarity- mechanical solidarity and organic solidarity. Mechanical solidarity prevails in simple folk societies where division of labour is restricted to family, village or small region. Here individuals do not differ much from one other and follow the same set of norms, beliefs etc. Organic solidarity holds the modern societies together with a bond. Here societies are large and people are engaged in variety of economic activities. They hold different values and socialize their children in varying patterns. The conditions of the modern society compel division of labour to reach the extreme level. This extreme form of division of labour leads to feeling of individualism or anomie. Anomie according to Durkheim refers to a state of normlessness in both the society and the individual. It is a social condition characterised by the breakdown of norms governing social interaction. People feel detached from their fellows having little commitment to shared norms people lack social guidelines for personal conduct. They are inclined to pursue their private interests without regard for the interests of society as a whole.

Karl Polyani: According to economist Karl Polanyi, the three principles of exchange are market principle, redistribution, and reciprocity. The market principle describes the buying and selling of goods and services based on the laws of supply and demand (things cost more the scarcer they are and the more people want them), and often involves bargaining. It is associated with industrial societies and involves a complex division of labour and central government. In redistribution, products move from the local level to a hierarchical centre, are reorganised, and sent back down to the local level. Redistribution is the main form of exchange in chiefdoms and some industrial states, and works with the market system. Polyani identifies reciprocity of three kinds: generalized, balanced, or negative. Generalized reciprocity involves an exchange between closely related people in which the giver expects nothing concrete or immediate in return. It is not necessarily classified as altruism, but resembles sharing by social contract. Generalized reciprocity is demonstrated by most egalitarian forager groups including the !Kung people, who do not say thank you upon receiving gifts because it is expected that at a later time, the act of goodwill will be reciprocated. It is also shown in most cases between parents and children. Another form of reciprocity is balanced reciprocity, in which the social distance between giver and recipient increases relative to generalized reciprocity. It involves an exchange outside the immediate family, and the giver expects something in return in the future, but not immediately. If there is no reciprocation, the relationship between the two parties will be strained. The third kind of reciprocity is negative reciprocity, which is an exchange relationship in which parties do not trust each other and are strangers. The giving must be reciprocated immediately and there is very little communication, if any, between groups. Each group is trying to maximize its economic benefit, but eventually friendly relationships between the groups may develop. An example of negative reciprocity is the Mbuti Pygmy foragers of Africa, who exchange with villagers in neighbouring groups in silent trade in which they place the items for exchange on the ground, then hide at a distance and wait for the other group to make an offer of their goods. Bartering may continue back and forth, but no direct contact is made between groups. Potlatching among the Kwakitul of Washington and British Columbia can be classified in the category of redistribution.

Economics

The term economics comes from Greek and refers to house holding (from oikos household and nomia a set of norms or laws). Aristotle used it in the sense of administration of domestic matters. Only in the 17th century its use extended from the private sphere to the public one with the work of the first mercantilist economists where the term came to refer to the study of the features and economic problems of a nation state. This started the discussion on political economy.

A society's economic system is the social institution that coordinates human activity to pro- duce, distribute, and consume goods and services. Goods include any product that is extracted from the earth, manufactured, or grown such as food, clothing, petroleum, natural gas, automobiles, coal, and computers. Services include activities performed for others such as entertainment, transportation, financial advice, medical care, spiritual counseling, and education.

The term political economy would continue to be used in Smith's Wealth of Nations in 1776. Only later towards the end of the 18th century would the term economics as against political economy come into use. This was introduced in Britain and USA through Alfred Marshall's Principle of Economics in 1890 that tended to underline the autonomy of economics from politics and the distinction between scientific analysis and economic policy.

In Adam Smith's view economy and society and market and institutions were still closely related. He clearly believed that the pursuit of individual interest and the workings of the market would only conduct to the general good when they are constrained by precise institutional rules. Smith saw the study of these institutional constraints as an integral part of research into the causes of the wealth of nations. Economics and economic sociology were closely linked in his work. Adam Smith published 'An inquiry into the nature and causes of the wealth of nations' in 1776. This text was basic to the formation of economics as a scientific discipline where he presents systematically presents laws on the functioning of the market in the production of goods and distribution of goods and the distribution of incomes. The book also examines the causes of wealth of nations or economic growth. The role of institutions is viewed as a given where as in dynamic part they are presented as a variable that can depending on their characteristics favour or discourage development.

Smith's work on static equilibrium presupposed that new wealth was not created but rather that existing wealth was used to satisfy needs. Capitalist institutions were taken as given. The land belonged to owners who lived on the profits and that workers depended on the sale of their work for a wage. He supposed that there were many sellers that information circulated freely and that the resources of capital and labour could be easily transferred from one use to another. In the institutional context the quantity of goods produced will tend to correspond to the demand that exists for such goods.

In the final analysis consumer demand will determine what is produced given the costs of production for each good. If the quantity of certain goods offered falls below demand price would increase as competition between buyers who wish to obtain the good drives them to pay more. As per the distribution of income it was also assumed that there was a particular market price for wages, profits and incomes. Wages tended to rise if the demand for labour was greater than supply and vice versa. Smith described a situation in which members of society were not rewarded for their participation in economic activity through traditional means of reciprocity or through legal regulation guaranteed by the power of the state redistribution as in the feudal system. Instead remuneration depended on the market and its laws.

The world's two major economic systems are capitalism and socialism. In capitalism, private citizens own the means of production and pursue profits. In socialism, the state owns the means of production and has no goal of profit. Adherents of each have developed ideologies that defend their own systems and paint the other as harmful or even evil.

Industrial Society

The steam engine invented in 1765, ushered in industrial societies. Based on machines powered by fuels, these societies created a surplus that stimulated trade among nations.

Then came more efficient machines. As the surpluses grew even greater, the emphasis gradually changed from producing goods to consuming them. In 1912, sociologist Thorstein Veblen coined the term conspicuous consumption to describe this fundamental change in people's orientations.

Veblen meant that the Protestant ethic identified by Weber an emphasis on hard work, savings and a concern for salvation was being replaced by an eagerness to show off wealth by the elaborate consumption of goods.

Postindustrial Societies

In 1973, sociologist Daniel Bell noted that a new type of society was emerging. He described the essential changes that are accompanying the emergence of a post-industrial society, one that relies on intellectual technologies of telecommunications and computers, not just "large computers but computers on a chip"

This new postindustrial society has six characteristics:

(1) A service sector so large that most people work in it,

(2) A vast surplus of goods,

(3) Even more extensive trade among nations

(4) A wider variety and quantity of goods available to the average person

(5) An information explosion

(6) A global village where the world's nations are linked by fast communications, transportation and trade.

In addition to the associated technology, a substantial proportion of the working population employed in service, sales, and administrative support occupations distinguishes post-industrial societies. There is an extraordinary rise in the percentage of workers in management, professional, and related occupations. There is an increased emphasis on education as the avenue of social mobility. This has led to an opportunity, the dominance of intellectual technology based on mathematics and linguistics in the form of algorithms, programs (software), models, and simulations the creation of an electronically mediated global communication infrastructure, which includes broadband, cable, digital TV, optical fibre networks, fax, e-mail and ISDN (integrated system digital networks) an economy defined not simply by the production of goods and labour-saving devices but by applied knowledge as the source of invention and innovation and by the manipulation of numbers, words, images, and other symbols.

The jobs associated with this knowledge driven, information based economy include computer programmers, technical writers, financial analysts, market analysts, and customer-service representatives. The challenge of post-industrial society is interpersonal, as the "basic experience of each person's life is his relationship between himself and others." In an environment that emphasizes knowledge and interpersonal relationships, the institutions of science and education take centre stage. With regard to science, Bell described the rise and importance of science-based industries, which involve applications of theoretical knowledge. These industries are fundamentally different from the industries of the Industrial Revolution, such as steel, automobile, and telephone. For the most part, these industries were "founded or created by talented tinkers" who were not connected to the scientific establishment.

Post-industrial industries derive directly from the investigations of scientists into the basic phenomena of nature and the applications of this research to technological problems. Education becomes key to negotiating an information society and is viewed as something that takes place across the lifespan, not con- fined to a specific time or place.

Medium of Exchange

A medium of exchange is the means by which people value and exchange goods and services. Hunting and gathering and pastoral and horticultural societies produced little surplus, and people bartered, directly exchanging one item for another. In later societies, surpluses grew and trade expanded people then developed new ways of placing values on goods and services so they could trade them.

Although bartering continued in agricultural societies, people increasingly came to use money, a medium of exchange that places a value on items. In most places, money consisted of gold and silver coins. A coin's weight and purity determined the amount of goods or services it could purchase.

Toward the end of the agricultural period, currency (paper money) came into existence. Each piece of paper represented a specific amount of gold or silver stored in a ware- house. Currency represented stored value. Gold and silver coins continued to circulate alongside the deposit receipts and currency.

When fiat money replaced stored value, coins made of precious metals disappeared from circulation. People considered these coins more valuable, and they were unwilling to part with them. Then as inferior metals (copper, zinc, and nickel) replaced the smaller silver coins, people began to hoard these silver coins, and they, too, disappeared from circulation. Even without a gold standard that restricts the amount of currency issued to the amount of stored value, governments have a practical limit on the amount of paper money they can distribute.

In general, prices increase if a government issues currency at a rate higher than the growth of its gross domestic product (GDP), the total goods and services that a country produces. This condition inflation means that each unit of currency will purchase fewer goods and services. Governments try to control inflation, for high inflation is a destabilizing influence. During the first part of the postindustrial society, paper money circulated freely. Paper money then became less common, gradually being replaced by checks and credit cards. The next development was the debit card; a device that electronically withdraws the cost of an item from the cardholder's bank account. Like the check, the debit card is a type of deposit receipt, for it transfers ownership of currency on deposit.

The latest evolution of money is e-cash, money stored on a company's computer that can be transferred over the Internet to anyone who has an account with that company. E-cash can be used for making purchases and for paying bills. We can use the term e-currency to refer to the most common form of e-cash. E-currency represents an amount recorded in a government's paper money, such as so many dollars or euros. The second form of e-cash is electronic gold, which represents a balance in units of gold. A transaction in e-gold is actually the transfer of ownership to a specified amount of gold that the owner has stored in a bank's vault.

Globalization of Capitalism

The globalization of capitalism may be the most significant economic change in the past 100 years. According to Louis Gallambos this new global business system will change the way everyone lives and works. From the functionalist perspective, work is a basis of social solidarity. According to Emile Durkheim as the farmers do the same type of work; they share a similar view of the world. He used the term mechanical solidarity to refer to the sense of unity that comes from doing similar activities. When an agricultural society industrializes, people work at many different types of jobs. As the division of labour grows, people come to feel less solidarity with one another. As they are like the separate organs that make up the same body, Durkheim called this type of unity organic solidarity.

This process has continued to the point that we now are developing a global division of labour as each of us now depends on workers around the globe. Corporations, with their separation of ownership and management, underlie the success of capitalism. We may not feel a sense of unity with one another, but the same global economic web links all of us. The globalization of capitalism has forged a new world structure. Three primary trading blocs have emerged: North and South America, dominated by the United States; Europe, dominated by Germany; and Asia, dominated by Japan and China. Functionalists stress that this new global division benefits not only the multinational giants but also the citizens of the world.

That capitalism could become the world's dominant economic force can be traced to a social invention called the corporation. A corporation is a business that is treated legally as a person. A corporation can make contracts, incur debts, sue and be sued. Its liabilities and obligations, however, are separate from those of its owners. One of the aspects of corporations is their separation of ownership and management. It is not the owners those who own the company's stock who run the day-to-day affairs of the company instead, managers run the corporation. The result is the "ownership of wealth without appreciable control, and control of wealth without appreciable ownership"

Conflict theorists stress how power is concentrated in the capitalist class. They note that global capitalism is a means by which capitalists exploit workers. From the major owners of the multinational corporations comes an inner circle. While workers lose jobs to automation, the inner circle maintains its political power and profits from the new technology. The term corporate capitalism indicates that giant corporations dominate capitalism today. Power and wealth have become so concentrated that a global superclass has arisen.

A tool that unites and magnifies their power is interlocking directorates. Individuals serve on the board of directors of several major companies, and so do their fellow board members. Like a spider's web that starts at the centre and fans out in all directions, these overlapping memberships join the top companies into a single network. The overlapping memberships of the globe's top multinational companies enfold their leaders into a small circle that are called the global superclass. The superclass is not only extremely wealthy but it is also extremely powerful. These people have access to the top circles of political power around the globe.

Industrial and Urban Society

Urbanization is a universal process implying economic development and social change. Urbanization also means a breakdown of traditional social institutions and values. However in India one cannot say that urbanization has resulted in the caste system being transformed into the class system, the joint family transforming into the nuclear family and religion becoming secularized. MSA Rao observes that the breakdown hypothesis originated from the western experience and it ignores the fact of traditional urbanization in India. Rao classifies urban studies in three categories â€" those concerned with the institutional approaches â€"those treating cities and their growth in the general context of history of civilization and those which formulate the cultural role of cities in the context of social organisation of the great tradition. The first category of studies highlights on economic institutions such as emergence of middle class and a commercial organisation and religion.

For Pirenne the city consisted of middle classes and groups engaged in trade and commerce. For Coulanges an ancient city was a religious community. Max Weber's emphasis was on social action and autonomous city government. The institutionalists look for specific causes and conditions for the growth of cities in different contexts. Robert Redfield has provided a typology of the city in terms of orthogenetic and heterogenetic processes of change in the organisation of tradition and culture. Milton Singer observes that the great tradition is basically an urban phenomenon and transformation of the little tradition into the great tradition refers to the process of urbanization. However great tradition has also been undergoing a significant change hence individualism, freedom and fluidity in traditional norms and values. Gideon Sjoberg distinguishes cities into pre-industrial and industrial. The preindustrial city was a feudal one. There are two limitations to this approach- feudalism was not the only basis of city formation and today the modern city is found in existence due to other factors too in addition to industrialization.

A number of criteria have been used to understand urban social structure and stratification. The most important ones are the extent of closure or openness and the nature of deprivations and gratifications. These apply to specific groups and collectivities as some of them have opportunities for betterment of their social standing whereas others remain deprived of the same. The individual is the basis in regard to motivational structure, utilisation of available opportunities and use of means of communication for realising one's aspirations. Urban social structure can be characterized in terms of having openness, attributional criteria, mobility and individual ranking.

Victor D'Souza has analyzed kinship, caste, class, religion and displaced or non-placed conditions in his study of the City of Chandigarh. Internal differentiation among different groupings has been analyzed on the basis of education, occupational prestige and income. The assumption is that if the groupings of a particular type are alike in respect of education, occupation and income then the principle on which they are formed is not an important basis of social organisation. The educational, occupational and income hierarchies are significantly correlated with each other. However the correlation of each of them with the operational caste hierarchy is not significant. Social class position is positively correlated with education and family income.

Cities consist of a variety of professional classes. They perform specialised functions such as teaching; medical, legal etc. study of the social origins of professionals may offer significant insights into the process of social stratification and mobility. Compared to other Asian countries the professional classes in India constitute a very small proportion of all workers. The upper, upper middle and middle classes dominate most of the opportunities and positions in the professions. The upper and upper middle caste and class background of most of the top-level political elite conforms to this general pattern. However in recent years the rural rich are replacing the urban rich in the field of politics to a large extent. Some change is also visible in the recruitment to various civil services and medical and engineering professions. The social structure of town and cities comprises of top-level businessmen, industrialists and bureaucrats, higher income professionals, scientists, technicians, professional managers in industry and large merchants. Clerks and minor officials in government offices and private firms, school teachers, petty shopkeepers and entrepreneurs and members of working class such as operators, artisans, household industry workers, service workers, hawkers, peddlers, construction workers and unskilled workers.

Urbanization is a worldwide phenomenon; India has also witnessed an increased growth of urbanization and industrialization in the post independence period. Urban growth in India is particularly due to large-scale migration from villages to towns and cities as the latter offers better facilities for education and training and more and better avenues for employment. Towns and cities are an important factor in development of the region in which they are located. The development of towns and cities depends upon the support villages in the vicinity extend to them in terms of migration, supply of farm produce to the cities and purchase of consumer goods from them.

Rural - Urban Continuum

Some sociologists have used the concept of rural-urban continuum to stress the idea that there are no sharp breaking points to be found in the degree or quantity of rural urban differences. Robert Redfield has given the concept of rural -urban continuum on the basis of his study of Mexican peasants of Tepoztlain. The rapid process of urbanization through the establishment of industries, urban traits and facilities have decreased the differences between villages and cities. There are some sociologists whose treat rural-urban as dichotomous categories have differentiated the two at various levels including occupational differences, environmental differences, differences in the sizes of communities, differences in the density of population, differences in social mobility and direction of migration, differences in social stratification and in the systems of social interaction.

A third view regarding rural and urban communities has been given by Pocock who believe that both village and city are elements of the same civilization and hence neither rural urban dichotomy, nor continuum is meaningful. M.S.A. Rao points out in the Indian context that although both village and town formed part of the same civilization characterized by institution of kinship and caste system in pre-British India, there were certain specific institutional forms and organisational ways distinguishing social and cultural life in towns form that in village. Thus, according to Rao, Rural Urban continuum makes more sense

Ghurye believes that urbanization is migration of people from village to city and the impact it has on the migrants and their families.

Maclver remarks that though the communities are normally divided into rural and urban the line of demarcation is not always clear between these two types of communities. There is no sharp demarcation to tell where the city ends and country begins. Every village possesses some elements of the city and every city carries some features of the village.

R.K Mukherjee prefers the continuum model by talking of the degree of urbanization as a useful conceptual tool for understanding rural-urban relations.

P.A Sorokin and Zimmerman in 'Principles of Rural-Urban sociology have stated that the factors distinguishing rural from urban communities include occupation, size and density of population as well as mobility, differentiation and stratification.

Urban Growth and Urbanization

Urbanization is the movement of population from rural to urban areas and the resulting increasing proportion of a population that resides in urban rather than rural places. It is derived from the Latin 'Urbs' a term used by the Romans to a city. Urban sociology is the sociology of urban living; of people in groups and social relationship in urban social circumstances and situation. Thompson Warren has defined it as the movement of people from communities concerned chiefly or solely with agriculture to other communities generally larger whose activities are primarily centred in government, trade, manufacture or allied interests. Urbanization is a two-way process because it involves not only movement from village to cities and change from agricultural occupation to business, trade, service and profession but it also involves change in the migrants attitudes, beliefs, values and behaviour patterns. The process of urbanization is rapid all over the world. The facilities like education, healthcare system, employment avenues, civic facilities and social welfare are reasons attracting people to urban areas. The census of India defines some criteria for urbanization. These are:

- Population is more than 5000
- The density is over 400 persons per sq.km
- 75% of the male population engages in non-agricultural occupations.
- Cities are urban areas with population more than one lakh.
- Metropolises are cities with population of more than one million.

Urbanism

Urbanism is a way of life. It reflects an organisation of society in terms of a complex division of labour, high levels of technology, high mobility, interdependence of its members in fulfilling economic functions and impersonality in social relations. Louis Wirth has given four characteristics of urbanism

- Transiency: An urban inhabitant's relation with others last only for a short time; he tends to forget his old acquaintances and develop relations with new people. Since he is not much attached to his neighbours members of the social groups, he does not mind leaving them.
- Superficiality: An urban person has the limited number of persons with whom he interacts and his relations with them

are impersonal and formal. People meet each other in highly segmental roles. They are dependent on more people for the satisfaction of their life needs.

- Anonymity: Urbanities do not know each other intimately. Personal mutual acquaintance between the inhabitants which ordinarily is found in a neighbourhood is lacking.
- Individualism: People give more importance to their own vested interests.

Town

The town is intermediate between rural and urban communities. It is too large for all inhabitants to be acquainted with one another, yet small enough for informal relationships to predominate. Social behaviour more closely resembles the rural than the metropolitan city pattern. Towns are places with population of 5,000 and more. Three conditions of a place being classified as a town are:

- The population is more than 5,000.
- The density is not less than 400 sq.km.
- Not less than 75% of the adult male population is engaged in non -agricultural activities.

City

Cities become possible when an agricultural surplus develops together with improved means of transportation and tend to be located at breaks in transportation. The most significant current developments in city structure are the metropolitan area including the suburb which accounts for current population growth. The city pulls people from various corners towards its nucleus. The rural people faced with various economic problems are attracted by the city and start moving towards the cities. The city provides ample opportunities for personal advancement. It is the centre of brisk economic, commercial, artistic, literary, political, educational, technological, scientific and other activities. Cities are not only the controlling centres of their societies but also the source of innovation and change. They act as the source of new ideas for production, the pace -setters for consumption, guardians of culture and conservers of order in society. Consensus and continuity in a society are maintained from the city centres. Urban culture has become the legitimation for control.

Walter Christaller explained the location of urban cities in terms of their functions as service centres. The basic assumption was that a given rural area supports an urban centre which in turn serves the surrounding countryside. There are smaller towns for smaller areas and bigger cities for larger regions. This concept permitted Christaller to build up an integrated system of cities according to their size.

His views conceiving a city as a central place within a rural area was elaborated by Edward L.Ullman with considerable modifications. He admits the vulnerability of the scheme for larger places. In highly industrialized areas the central place schemes is generally distorted by industrial concentration in response to resources and transportation that it may be said to have little significance as an explanation for urban location and distribution.

Hyot in his sector theory talked about the growth of cities taking place in sectors and these sectors extend from the centre to periphery.

The concentric zone theory given by Park and Burgess suggested that modern cities consisted of a series of concentric zones. There are five such zones

- Central business district
- Zone in transition
- Zone of working population
- Residential zone
- Commuter's zone

Gans and Lewis through compositional theory hold that the composition of a city's population differs from that of a small town in terms of factors such as class, education, ethnicity and marital status.

Multiple Nuclie theory given by Harris and Ullman discuss that there is not one centre but several centres for the city. Each of the centres tend to specialise in a particular kind of activity-retailing, wholesaling, finance, recreation, education, government. Several centres may have existed from the beginning of the city or many have developed later in a division from one centre.

According to Castells to understand cities and urbanism one has to understand the process by which spatial forms are created and transformed. The architecture of cities expresses the struggles and conflicts between different groups in society. City is not only a distinct location but also as an integral part of processes of collective consumption.

Urban Ecological Processes

It means whereby spatial distribution of people and activities change. They include:

- Centralization clustering of economic and service functions.
- Concentration tendency of people and activities to cluster together.
- Decentralization flight of people and activities from the centre of the city.
- Invasion entrance of new kind of people or activity into an area.
- Segregation concentration of a certain type of people or activities within a particular area.
- Succession completed replacement of one kind of people or activity by another.

Impact of Automation on Society

- It speeds up the developmental processes of the society.
- It increases production.
- Brings further technological changes like information technology.
- Extreme industrialization
- Replacement of human labour with machines.
- Increase in profit margins
- Distance reduction through technological advancements in the field of communication network.
- Makes life dependent on latest gizmos and equipments.

Environment

When physical, chemical and biological projects of the different components of environment: air, water, soil, noise change to the detriment of living of humans it may be said that environment has been affected. Many developing countries are placing more and more reliance on industrialization. It is not only a mechanical but also a social process. Therefore it affects the environment physically as well as socio-culturally. All aspects of pollution are directly or indirectly related to human health and well being. The excessive growth and rush of people from villages to urban areas resulting in over crowding of cities. Rapid urbanization and industrialization have led to an increase in environmental pollutant load that poses serious public health problem. It also affects the socio-cultural environment with the close ties of groups coming under pressure. Traditional ties are replaced with new work based ones. Religion becomes secular. Thus industrialization affects the social fabric making the society more materialistic.

Social Inequality

Social inequality occurs when resources in a given society are distributed unevenly, typically through norms of allocation, that engender specific patterns along lines of socially-defined categories of persons. Economic inequality, usually described on the basis of the unequal distribution of income or wealth, is a frequently studied type of social inequality. Though the disciplines of economics and sociology generally use different theoretical approaches to examine and explain economic inequality, both fields are actively involved in researching this inequality. However, social and natural resources other than purely economic resources are also unevenly distributed in most societies and may contribute to social status. Norms of allocation can also affect the distribution of rights and privileges, social power, access to public goods such as education or the judicial system, adequate housing, transportation, credit and financial services such as banking and other social goods and services.

While many societies worldwide hold that their resources are distributed on the basis of merit, research shows that the distribution of resources often follows delineations that distinguish different social categories of persons on the basis of other socially-defined characteristics. For example, social inequality is linked to racial inequality, gender inequality, and ethnic inequality as well as other status characteristics.

Overview

Social inequality is found in almost every society. In simple societies, those that have few social roles and statuses occupied by its members, social inequality may be very low. In tribal societies, for example, a tribal head or chieftain may hold some privileges, use some tools, or wear marks of office to which others do not have access, but the daily life of the chieftain is very much like the daily life of any other tribal member. Anthropologists identify such highly egalitarian cultures as "kinship-oriented," which appear to value social harmony more than wealth or status. These cultures are contrasted with materially-oriented cultures in which status and wealth are prized and competition and conflict are common. Kinship-oriented cultures may actively work to prevent social hierarchies from developing because they believe that could lead to conflict and instability. In today's world, most of our population lives in more complex than simple societies. As social complexity increases, inequality tends to increase along with a widening gap between the poorest and the most wealthy members of society.

Social status is accorded to persons in a society on at least two bases: ascribed characteristics and achieved characteristics. Ascribed characteristics are those present at birth or assigned by others and over which an individual has little or no control. Examples include sex (male or female), skin colour, eye shape, place of birth, parentage and social status of parents. Achieved characteristics are those which we earn or choose; examples include level of education, marital status, leadership status and other measures of merit. In most societies, an individual's social status is a combination of ascribed and achieved factors. In some societies, however, only ascribed statuses are considered in determining one's social status and there exists little to no social mobility and, therefore, few paths to more social equality. This type of social inequality is generally referred to as caste inequality.

One's social location in a society's overall structure of social stratification affects and is affected by almost every aspect of social life and one's life chances. The single best predictor of an individual's future social status is the social status into which they were born. Theoretical approaches to explaining social inequality concentrate on questions about how such social differentiations arise, what types of resources are being allocated, what are the roles of human cooperation and conflict in allocating resources, and how do differing types and forms of inequality affect the overall functioning of a society?

The variables considered most important in explaining inequality and the manner in which those variables combine to produce the inequities and their social consequences in a given society can change across time and place. In addition to interest in comparing and contrasting social inequality at local and national levels, in the wake of today's globalizing processes, the most interesting question becomes: what does inequality look like on a worldwide scale and what does such global inequality bode for the future? In effect, globalization reduces the distances of time and space, producing a global interaction of cultures and societies and social roles that can increase global inequities.

Inequality and Ideology

Philosophical questions about social ethics and the desirability or inevitability of inequality in human societies have given rise to a spate of ideologies to address such questions. We can broadly classify these ideologies on the basis of whether they justify or legitimize inequality, casting it as desirable or inevitable, or whether they cast equality as desirable and inequality as a feature of society to be reduced or eliminated. One end of this ideological continuum can be called

"Individualist", the other "Collectivist". In Western societies, there is a long history associated with the idea of individual ownership of property and economic liberalism, the ideological belief in organising the economy on individualist lines such that the greatest possible number of economic decisions are made by individuals and not by collective institutions or organisations. Laissez-faire, free market ideologies—including classical liberalism, neoliberalism and libertarianism—are formed around the idea that social inequality is a "natural" feature of societies, is therefore inevitable and, in some philosophies, even desirable. Inequality provides for differing goods and services to be offered on the open market, spurs ambition, and provides incentive for industriousness and innovation. At he other end of the continuum, collectivists place little to no trust in "free market" economic systems, noting widespread lack of access among specific groups or classes of individuals to the costs of entry to the market. Widespread inequalities often lead to conflict and dissatisfaction with the current social order. Such ideologies include Fabianism, socialism, and Marxism or communism. Inequality, in these ideologies, must be reduced, eliminated, or kept under tight control through collective regulation.

Though the above discussion is limited to specific Western ideologies, it should be noted that similar thinking can be found, historically, in differing societies throughout the world. While, in general, eastern societies tend toward collectivism, elements of individualism and free market organisation can be found in certain regions and historical eras. Classic Chinese society in the Han and Tang dynasties, for example, while highly organised into tight hierarchies of horizontal inequality with a distinct power elite also had many elements of free trade among its various regions and subcultures.

Today, there is belief held by some that social inequality often creates political conflict and growing consensus that political structures determine the solution for such conflicts. Under this line of thinking, adequately designed social and political institutions are seen as ensuring the smooth functioning of economic markets such that there is political stability, which improves the long-term outlook, enhances labour and capital productivity and so stimulates economic growth. With higher economic growth, net gains are positive across all levels and political reforms are easier to sustain. This may explain why, over time, in more egalitarian societies fiscal performance is better, stimulating greater accumulation of capital and higher growth.

Inequality and Social Class

Socioeconomic status (SES) is a combined total measure of a person's work experience and of an individual's or family's economic and social position in relation to others, based on income, education, and occupation. It is often used as synonymous with social class, a set of hierarchical social categories that indicate an individual's or household's relative position in a stratified matrix of social relationships. Social class is delineated by a number of variables, some of which change across time and place. For Karl Marx, there exist two major social classes with significant inequality between the two. The two are delineated by their relationship to the means of production in a given society. Those two classes are defined as the owners of the means of production and those who sell their labour to the owners of the means of production. In capitalistic societies, the two classifications represent the opposing social interests of its members, capital gain for the capitalists and good wages for the laborers, creating social conflict.

Max Weber uses social classes to examine wealth and status. For him, social class is strongly associated with prestige and privileges. It may explain social reproduction, the tendency of social classes to remain stable across generations maintaining most of their inequalities as well. Such inequalities include differences in income, wealth, access to education, pension levels, social status, socioeconomic safety-net. In general, social class can be defined as a large category of similarly ranked people located in a hierarchy and distinguished from other large categories in the hierarchy by such traits as occupation, education, income, and wealth.

In modern Western societies, inequalities are often broadly classified into three major divisions of social class: upper class, middle class, and lower class. Each of these classes can be further subdivided into smaller classes (e.g. "upper middle"). Members of different classes have varied access to financial resources, which affects their placement in the social stratification system.

The quantitative variables most often used as an indicator of social inequality are income and wealth. In a given society, the distribution of individual or household accumulation of wealth tells us more about variation in well-being than does income, alone. Gross Domestic Product (GDP), especially *per capita* GDP, is sometimes used to describe economic inequality at the international or global level. A better measure at that level, however, is the Gini coefficient, a measure of statistical dispersion used to represent the distribution of a specific quantity, such as income or wealth, at a global level, among a nation's residents, or even within a metropolitan area. Other widely used measures of economic inequality are the percentage of people living with under USD$1.25 or USD$2 a day and the share of national income held by the wealthiest 10% of the population, sometimes called "the Palma" measure.

Patterns of Inequality

There are a number of socially-defined characteristics of individuals that contribute to social status and, therefore, equality or inequality within a society. When researchers use quantitative variables such as income or wealth to measure inequality, on an examination of the data, patterns are found that indicate these other social variables contribute to income or wealth as intervening variables. Significant inequalities in income and wealth are found when specific socially-defined categories of people are compared. Among the most pervasive of these variables are sex/gender, race, and ethnicity. This is not to say, in societies wherein merit is considered to be the primary factor determining one's place or rank in the social order, that merit has no effect on variations in income or wealth. It is to say that these other socially-defined characteristics can, and often do, intervene in the valuation of merit.

Gender Inequality

Sex- and gender-based prejudice and discrimination, called sexism, are major contributing factors to social inequality. Most societies, even agricultural ones, have some sexual division of labour and gender-based division of labour tends to increase during industrialization. The emphasis on gender inequality is born out of the deepening division in the roles assigned to men and women, particularly in the economic, political and educational spheres. Women are under-represented in political activities and decision making processes in most states in both the Global North and Global South.

Gender discrimination, especially concerning the lower social status of women, has been a topic of serious discussion not only within academic and activist communities but also by governmental agencies and international bodies such as the United Nations. These discussions seek to identify and remedy widespread, institutionalized barriers to access for women in their societies. By making use of gender analysis, researchers try to understand the social expectations, responsibilities, resources and priorities of women and men within a specific context, examining the social, economic and environmental factors which influence their roles and decision-making capacity. By enforcing artificial separations between the social and economic roles of men and women, the lives of women and girls are negatively impacted and this can have the effect of limiting social and economic development. Cultural ideals about women's work can also affect men whose outward gender expression is considered "feminine" within a given society. Transgender and gender-variant persons may express their gender through their appearance, the statements they make, or official documents they present. In this context, gender normativity, which is understood as the social expectations placed on us when we present particular bodies, produces widespread cultural/institutional devaluations of trans identities, homosexuality and femininity. Trans persons, in particular, have been defined as socially unproductive and disruptive.

A variety of global issues like HIV/AIDS, illiteracy, and poverty are often seen as "women's issues" since women are disproportionately affected. In many countries, women and girls face problems such as lack of access to education, which limit their opportunities to succeed, and further limits their ability to contribute economically to their society. Women are under-represented in political activities and decision making processes throughout most of the world. As of 2007, around 20 percent of women were below the $1.25/day international poverty line and 40 percent below the $2/day mark. More than one-quarter of females under the age of 25 were below the $1.25/day international poverty line and about half on less than $2/day.

Women's participation in work has been increasing globally, but women are still faced with wage discrepancies and differences compared to what men earn. This is true globally even in the agricultural and rural sector in developed as well as developing countries. Structural impediments to women's ability to pursue and advance in their chosen professions often result in a phenomenon known as the glass ceiling, which refers to unseen - and often unacknowledged barriers that prevent minorities and women from rising to the upper rungs of the corporate ladder, regardless of their qualifications or achievements. This effect can be seen in the corporate and bureaucratic environments of many countries, lowering the chances of women to excel. It prevents women from succeeding and making the maximum use of their potential, which is at a cost for women as well as the society's development. Ensuring that women's rights are protected and endorsed can promote a sense of belonging that motivates women to contribute to their society. Once able to work, women should be titled to the same job security and safe working environments as men. Until such safeguards are in place, women and girls will continue to experience not only barriers to work and opportunities to earn, but will continue to be the primary victims of discrimination, oppression, and gender based violence.

Women and persons whose gender identity does not conform to patriarchal beliefs about sex (only male and female) continue to face violence on global domestic, interpersonal, institutional and administrative scales. While first-wave Liberal Feminist initiatives raised awareness about the lack of fundamental rights and freedoms that women have access to, second-wave feminism highlighted the structural forces that underlie gender based violence. Masculinities are generally constructed so as to subordinate femininities and other expressions of gender that are not heterosexual, assertive and dominant. Gender sociologist and author, Raewyn Connell, discusses in her 2009 book, Gender, how masculinity is dangerous, heterosexual, violent and authoritative. These structures of masculinity ultimately contribute to the vast amounts of gendered violence, marginalization and suppression that women, queer, transgender, gender variant and gender non-conforming persons face. Some scholars suggest that women's under-representation in political systems speaks the idea that "formal citizenship does not always imply full social membership". Men, male bodies and expressions of masculinity are linked to ideas about work and citizenship. Others point out that patriarchal states tend top scale and claw back their social policies relative to the disadvantage of women. This process ensures that women encounter resistance into meaningful positions of power in institutions, administrations, and political systems and communities.

Racial and Ethnic Inequality

Racial or ethnic inequality is the result of hierarchical social distinctions between racial and ethnic categories within a society and often established based on characteristics such as skin colour and other physical characteristics or an individual's place of origin or culture. Even though race has no biological connection, it has become a socially constructed category capable of restricting or enabling social status. Unequal treatment and opportunities between such categories is usually the result of some categories being considered superior to others. This inequality can manifest through discriminatory hiring and pay practices. In some cases, employers have been shown to prefer hiring potential employees based on the perceived ethnicity of a candidate's given name - even if all they have to go by in their decision are resumes featuring identical qualifications. These sorts of discriminatory practices stem from prejudice and stereotyping, which occurs when people form assumptions about the tendencies and characteristics of certain social categories, often rooted in assumptions about biology, cognitive capabilities, or even inherent moral failings. These negative attributions are then disseminated through a society through a number of different mediums, including television, newspapers and the internet, all of which play a role in promoting preconceived notions of race that disadvantage and marginalize groups of people. This along with xenophobia and other forms of discrimination continue to occur in societies with the rise of globalization.

Racial inequality can also result in diminished opportunities for members of marginalized groups, which in turn can lead to cycles of poverty and political marginalization. Racial and ethnic categories become a minority category in a society. Minority members in such a society are often subjected to discriminatory actions resulting from majority policies, including assimilation, exclusion, oppression, expulsion, and extermination. For example, during the run-up to the 2012 federal elections in the United States, legislation in certain "battleground states" that claimed to target voter fraud had the effect of disenfranchising tens of thousands of primarily African American voters. These types of institutional barriers to full and equal social participation have far-reaching effects within marginalized communities, including reduced economic opportunity and output, reduced educational outcomes and opportunities and reduced levels of overall health.

In the United States, research demonstrates that mass incarceration has been a modern tool of the state to impose inequality, repression, and discrimination upon African American and Hispanics. The War on Drugs has been a campaign with disparate effects, ensuring the constant incarceration of poor, vulnerable, and marginalized populations in North America. Over a million African Americans are incarcerated in the US, many of whom have been convicted of a drug possession charge. With the States of Colorado and Washington having legalized the possession of marijuana, drug reformists and anti-war on drugs lobbyists are hopeful that drug issues will be interpreted and dealt with from a healthcare perspective instead of a matter of criminal law. In Canada, Aboriginal, First Nations and Indigenous persons represent over a quarter of the federal prison population, even though they only represent 3% of the country's population.

Age Inequality

Age discrimination is defined as the unfair treatment of people with regard to promotions, recruitment, resources, or privileges because of their age. It is also known as ageism: the stereotyping of and discrimination against individuals or groups based upon their age. It is a set of beliefs, attitudes, norms, and values used to justify age-based prejudice, discrimination, and subordination. One form of ageism is adultism, which is the discrimination against children and people under the legal adult age. An example of an act of adultism might be the policy of a certain establishment, restaurant, or place of business to not allow those under the legal adult age to enter their premises after a certain time or at all. While some people may benefit or enjoy these practices, some find them offencive and discriminatory.

As implied in the definitions above, treating people differently based upon their age is not necessarily discrimination. Virtually every society has age-stratification, meaning that the age structure in a society changes as people begin to live longer and the population becomes older. In most cultures, there are different social role expectations for people of different ages to perform. Every society manages people's aging by allocating certain roles for different age groups. Age discrimination primarily occurs when age is used as an unfair criterion for allocating more or less resources. Scholars of age inequality have suggested that certain social organisations favour particular age inequalities. For instance, because of their emphasis on training and maintaining productive citizens, modern capitalist societies may dedicate disproportionate resources to training the young and maintaining the middle-aged worker to the detriment of the elderly and the retired (especially those already disadvantaged by income/wealth inequality).

In modern, technologically advanced societies, there is a tendency for both the young and the old to be relatively disadvantaged. However, more recently, in the United States the tendency is for the young to be most disadvantaged. For example, poverty levels in the U.S. have been decreasing among people aged 65 and older since the early 1970s whereas the number children under 18 in poverty has steadily risen. Sometimes, the elderly have had the opportunity to build their wealth throughout their lives, younger people have the disadvantage of recently entering into or having not yet entered into the economic sphere. The larger contributor to this, however is the increase in the number of people over 65 receiving Social Security and Medicare benefits in the U.S.

When we compare income distribution among youth across the globe, we find that about half (48.5 percent) of the world's young people are confined to the bottom two income brackets as of 2007. This means that, out of the three billion persons under the age of 24 in the world as of 2007, approximately 1.5 billion were living in situations in which they and their families had access to just nine percent of global income.

Moving up the income distribution ladder, children and youth do not fare much better: more than two-thirds of the world's youth have access to less than 20 percent of global wealth, with 86 percent of all young people living on about one-third of world income. For the just over 400 million youth who are fortunate enough to rank among families or situations at the top of the income distribution, however, opportunities improve greatly with more than 60 percent of global income within their reach.

Although this does not exhaust the scope of age discrimination, in modern societies it is often discussed primarily with regards to the work environment. Indeed, nonparticipation in the labour force and the unequal access to rewarding jobs means that the elderly and the young are often subject to unfair disadvantages because of their age. On the one hand, the elderly are less likely to be involved in the workforce: At the same time, old age may or may not put one at a disadvantage in accessing positions of prestige. Old age may benefit one in such positions, but it may also disadvantage one because of negative ageist stereotyping of old people. On the other hand, young people are often disadvantaged from accessing prestigious or relatively rewarding jobs, because of their recent entry to the work force or because they are still completing their education. Typically, once they enter the labour force or take a part-time job while in school, they start at entry level positions with low level wages. Furthermore, because of their lack of prior work experience, they can also often be forced to take marginal jobs, where they can be taken advantage of by their employers.

Inequalities in Health

Health inequalities can be defined as differences in health status or in the distribution of health determinants between different population groups. Health inequalities are in many cases related to access to health care. In industrialized nations, health inequalities are most prevalent in countries that have not implemented a universal health care system, such as the United States. Because the US health care system is heavily privatized, access to health care is dependent upon one's economic capital; Health care is not a right, it is a commodity that can be purchased through private insurance companies (or that is sometimes provided through an employer). The way health care is organised in the U.S. contributes to health inequalities based on gender, socioeconomic status and race/ethnicity. As Wright and Perry assert, "social status differences in health care are a primary mechanism of health inequalities". In the United States, over 48 million people are without medical care coverage. This means that almost one sixth of the population is without health insurance, mostly people belonging to the lower classes of society.

While universal access to health care may not completely eliminate health inequalities, it has been shown that it greatly reduces them. In this context, privatization gives individuals the 'power' to purchase their own health care (through private health insurance companies), but this leads to social inequality by only allowing people who have economic resources to access health care. Citizens are seen as consumers who have a 'choice' to buy the best health care they can afford; in alignment with neoliberal ideology, this puts the burden on the individual rather than the government or the community.

In countries that have a universal health care system, health inequalities have been reduced. In Canada, for example, equity in the availability of health services has been improved dramatically through Medicare. People don't have to worry about how they will pay health care, or rely on emergency rooms for care, since health care is provided for the entire population. However, inequality issues still remain. For example, not everyone has the same level of access to services. Inequalities in health are not, however, only related to access to health care. Even if everyone had the same level of access, inequalities may still remain. This is because health status is a product of more than just how much medical care people have available to them. While Medicare has equalized access to health care by removing the need for direct payments at the time of services, which improved the health of low status people, inequities in health are still prevalent in Canada. This may be due to the state of the current social system, which bear other types of inequalities such as economic, racial and gender inequality.

A lack of health equity is also evident in the developing world, where the importance of equitable access to healthcare has been cited as crucial to achieving many of the Millennium Development Goals. Health inequalities can vary greatly depending on the country one is looking at. Inequalities in health are often associated with socioeconomic status and access to health care. Health inequities can occur when the distribution of public health services is unequal. For example, in Indonesia in 1990, only 12% of government spending for health was for services consumed by the poorest 20% of households, while the wealthiest 20% consumed 29% of the government subsidy in the health sector. Access to health care is heavily influenced by socioeconomic status as well, as wealthier population groups have a higher probability of obtaining care when they need it. A study by Makinen et al. (2000) found that in the majority of developing countries they looked at, there was an upward trend by quintile in health care use for those reporting illness. Wealthier groups are also more likely to be seen by doctors and to receive medicine.

Global Inequality

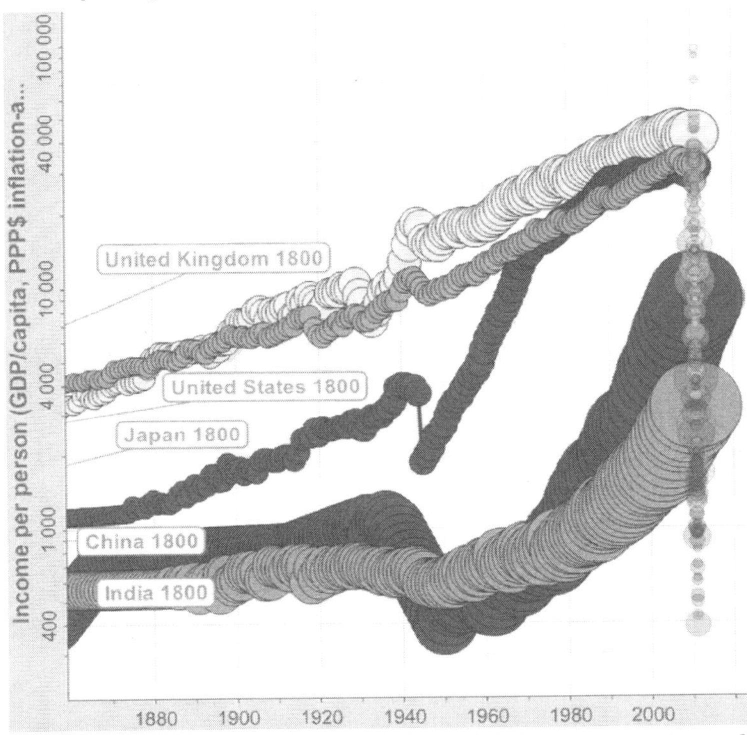

Figure: *Gross domestic product in 2011 US dollars per capita, adjusted for inflation and purchasing power parity (log scale) from 1860 to 2011, with population (disk area) for the US (yellow), UK (orange), Japan (red), China (red), and India (blue).*

The economies of the world have developed unevenly, historically, such that entire geographical regions were left mired in poverty and disease while others began to reduce poverty and disease on a wholesale basis. This was represented by a type of North–South divide that existed after WWII between First world, more developed, industrialized, wealthy countries and Third world countries, primarily as measured by GDP. From around 1980, however, through at least 2011, the GDP gap, while still wide, appeared to be closing and, in some more rapidly developing countries, life expectancies began to rise. However, there are numerous limitations of GDP as an economic indicator of social "well-being."

If we look at the Gini coefficient for world income, over time, after WW II the global Gini coefficient sat at just under .45. Between around 1959 to 1966, the global Gini increased sharply, to a peak of around .48 in 1966. After falling and leveling off a couple of times during a period from around 1967 to 1984, the Gini began to climb again in the mid-eighties until reaching a high or around .54 in 2000 then jumped again to around .70 in 2002. Since the late 1980s, the gap between some regions has markedly narrowed— between Asia and the advanced economies of the West, for example—but huge gaps remain globally. Overall equality across humanity, considered as individuals, has improved very little. Within the decade between 2003 and 2013, income inequality grew even in traditionally egalitarian countries like Germany, Sweden and Denmark. With a few exceptions—France, Japan, Spain—the top 10 percent of earners in most advanced economies raced ahead, while the bottom 10 percent fell further behind. By 2013, a tiny elite of multibillionaires, 85 to be exact, had amassed wealth equivalent to all the wealth owned by the poorest half (3.5 billion) of the world's total population of 7 billion. Country of citizenship (an ascribed status characteristic) explains 60% of variability in global income; citizenship and parental income class (both ascribed status characteristics) combined explain more than 80% of income variability.

Inequality and Economic Growth

The concept of economic growth is fundamental in capitalist economies. Productivity must grow as population grows and capital must grow to feed into increased productivity. Investment of capital leads to returns on investment (ROI) and increased capital accumulation. The hypothesis that economic inequality is a necessary precondition for economic growth has been a mainstay of liberal economic theory. Recent research, particularly over the first two decades of the 21st century, has called this basic assumption into question. While growing inequality does have a positive correlation with economic growth under specific sets of conditions, inequality in general is not positively correlated with economic growth and, under some conditions, shows a negative correlation with economic growth.

Milanovic (2011) points out that overall, global inequality between countries is more important to growth of the world economy than inequality within countries. While global economic growth may be a policy priority, recent evidence about regional and national inequalities cannot be dismissed when more local economic growth is a policy objective. The recent financial crisis and Global Recession hit countries and shook financial systems all over the world. This led to the implementation of large scale fiscal expansionary interventions and, as a result, to massive public debt issuance in some countries. Governmental bailouts of the banking system further burdened fiscal balances and raises considerable concern about the fiscal solvency of some countries. Most governments want to keep deficits under control but rolling back the expansionary measures or cutting spending and raising taxes implies an enormous wealth transfer from tax payers to the private financial sector. Expansionary fiscal policies shift resources and causes worries about growing inequality within countries. Moreover, recent data confirm an on-going trend of increasing income inequality since the early nineties. Increasing inequality within countries has been accompanied by a redistribution of economic resources between developed economies and emerging markets. Davtyn et al. (2014) studied the interaction of these fiscal conditions and changes in fiscal and economic policies with income inequality in the UK, Canada, and the USA. They find income inequality has negative effect on economic growth in the case of the UK but a positive effect in the cases of the USA and Canada. Income inequality generally reduces government net lending/borrowing for all the countries. Economic growth, they find, leads to an increase of income inequality in the case of the UK and to the decline of inequality in the cases of the USA and Canada. At the same time, economic growth improves government net lending/borrowing in all the countries. Government spending leads to the decline in inequality in the UK but to its increase in the USA and Canada.

Following the results of Alesina and Rodrick (1994), Bourguignon (2004) Birdsall (2005), and others, show that developing countries with high inequality tend to grow more slowly, Ortiz and Cummings (2011) show that developing countries with high inequality tend to grow more slowly. For 131 countries for which they could estimate the change in Gini index values between 1990 and 2008, they find that those countries that increased levels of inequality experienced slower annual per capita GDP growth over the same time period. Noting a lack of data for national wealth, they build an index using Forbes list of billionaires by country normalized by GDP and validated through correlation with a Gini coefficient for wealth and the share of wealth going to the top decile. They find that many countries generating low rates of economic growth are also characterized by a high level of wealth inequality with wealth concentration among a class of entrenched elites. They conclude that extreme inequality in the distribution of wealth globally, regionally and nationally, coupled with the negative effects of higher levels of income disparities, should make us question current economic development approaches and examine the need to place equity at the centre of the development agenda. Ostry et al.(2014) decisively reject the hypothesis that there is a major trade-off between a reduction of income inequality (through income redistribution) and economic growth. If that were the case, they hold, then redistribution that reduces income inequality would on average be bad for growth, taking into account both the direct effect of higher redistribution and the effect of the resulting lower inequality. Their research shows rather the opposite: increasing income inequality always has a significant and, in most cases, negative effect on economic growth while redistribution has an overall pro-growth effect (in one sample) or no growth effect. Their conclusion is that increasing inequality, particularly when inequality is already high, results in low growth, if any, and such growth may be unsustainable over long periods.

Piketty and Saez (2014) note that there are important differences between income and wealth inequality dynamics. First, wealth concentration is always much higher than income concentration. The top 10 percent of wealth share typically falls in the 60 to 90 percent range of all wealth, whereas the top 10 percent income share is in the 30 to 50 percent range. The bottom 50 percent wealth share is always less than 5 percent, whereas the bottom 50 percent income share generally falls in the 20 to 30 percent range. The bottom half of the population hardly owns any wealth, but it does earn appreciable income: The inequality of labour income can be high, but it is usually much less extreme. On average, members of the bottom half of the population, in terms of wealth, own less than one-tenth of the average wealth. The inequality of labour income can be high, but it is usually much less extreme. Members of the bottom half of the population in income earn about half the average income. In sum, the concentration of capital ownership is always extreme, so that the very notion of capital is fairly abstract for large segments—if not the majority—of the population. Piketty (2014) finds that wealth-income ratios, today, seem to be returning to very high levels in low economic growth countries, similar to what he calls the "classic patrimonial" wealth-based societies of the 19th century wherein a minority lives off its wealth while the rest of the population works for subsistence living. He surmises that wealth accumulation is high because growth is low.

Geography and Affect

Geographies of Affect: As Thrift (2004: 59) notes, "there is no stable definition of affect". The concept has been present in a number of different traditions and therefore has been understood in a number of senses. Very generally though, affect can be seen to refer to the process of transition through which a body goes; it is a "transpersonal *capacity* which a body has to be affected (through an affection) and to affect (as the result of modifications)" (Anderson 2006: 735). The emphasis here is on a "processual logic of *transitions* that take place during spatially and temporally distributed encounters in which 'each transition is accompanied by a variation in the capacity…'" (Massumi cited in Anderson 2006: 735). Thinking through affect implies that "the world is made up of billions of happy or unhappy encounters, encounters which describe a 'mindful connected physicalism' consisting of multitudinous paths which intersect" (Thrift 1999: 302). Put simply, an affect "is a mixture of two bodies, one body which is said to act on another, and the other receives traces of the first" (Deleuze 1978).

From this a body here has two simultaneous definitions. First, it is defined kinetically as being a composition of an infinite number of particles being at varying degrees of motion and rest, speed and slowness (longitude). Second, a body is defined dynamically by its capacity of affecting and being affected (latitude). This constructs an ecological map of the body, an immanent plan(e) "which is always variable and is constantly being altered, composed and recomposed, by individuals and collectives" (Deleuze 1988: 128). Deleuze (1988) discusses the modification of the body as being conditioned by two fundamental affects: joy and sadness. Joy refers to a positive affection, a nutrition, an increased speed and motion, increasing our capacity to act. Sadness is a negative affection, a poisoning, a slowing down, that reduces our capacity to act. Affects are then becomings: "sometimes they weaken us in so far as they diminish our power to act and decompose our relationships (sadness), sometimes they make us stronger in so far as they increase our power and make us enter into a vast or superior individual (joy)" (Deleuze and Parnet 2006: 45).

This understanding of affect can be both tied to, and distinguished from, feeling. Affects occur between objects or entities, and these interactions or affects are *felt* as intensities in the body, or find "corporeal expression in bodily feelings" (Anderson 2006: 736), and in so being are manifest in an alteration in a body's capacity to act. We can understand affect as "a kind of vague but intense atmosphere" and feeling as "that atmosphere felt in the body" (McCormack 2008: 6). Further, and again, these can both be distinguished from emotions. Of significance here is the work of Massumi (2002: 28) who distinguishes emotion from affect in defining emotion as

> *"a subjective content, the sociolinguistic fixing of the quality of an experience which is from that point onward*

> *described as personal. Emotion is qualified intensity, the conventional, consensual point of insertion of intensity into semantically and semiotically formed progressions, into narritavizable action-reaction circuits, into function and meaning. It is intensity owned and recognised".*

Emotion is therefore related to an "already established field of discursively constituted categories in relation to which the felt intensity of experience is articulated" and therefore "conceives experience as always already meaningful" (McCormack 2003: 495). This restricts the movement of affective intensities which exist prior to such fixing and framing by reducing them to such framings of meaning and significance.

This emphasis on the affective has produced a number of debates. Therefore, I will address two of these here. These relate to 1) how we understand affect, emotion, and power, and 2) to the ways in which the body is conceived as always active and agentive.

Debating Affect

This first set of debates just mentioned can be divided into two interrelated themes. Firstly, it has been argued that certain aspects of the language through which affect has been talked about is suggestive of a distancing from the emotional and the personal and instead promotes a focus on the reasonable and the public (Thien 2005). However, I simply do not see how affect can be seen as public. Relational yes – it occurs in the in-between; it does not belong to a subject. But this does not make it *public*. In fact one of the immense problems posed by trying to study affective experience is finding means through which to convey something of this affective experience given that it occurs outside of the realm of subjective, reflective experience.

In terms of affect being rational, again, this is simply not the case. Affect is precisely about the *pre-rational*. It comes before any *rationalization* by a subject. While it has been argued that, in contrast, emotion is not about the rational, for me, in that such works appeals to a (tacit) humanistic universalist logic whereby such emotional experiences mean the same thing to different people (Anderson and Harrison 2006; McCormack 2006), this it fact would be more suggestive of a rationalization of experience than work taking affect as its main focus.

Secondly, it has been argued that work on affect is inattentive to issues of power and that it is universalistic (Tolia-Kelly 2006). While I am sympathetic to the trajectory of Tolia-Kelly's argument – the development of a non-universalizing understanding of affective capacities that pays attention to the ways in which such capacities are socialized – I cannot help but feel that there are flawed steps in her argument and that due to this it falls foul of the approach she wishes to sidestep. Tolia-Kelly picks up on Deleuze's laying out of a 'common plane of immanence' and suggests that this 'universalizing' conception does not pay attention to the ways in which collectives are differently capable of affecting and being affected due to their access to geopolitical power, amongst other things. To an extent this is true, but in a very specific way. While this is a *universal* plane of immanence, it is only the same for everyone in that it is *different* for everyone. Therefore, this pays attention to the specificity of each encounter and does not introduce a universalistic principle.

This means that Deleuze pays attention to the specificity of how each *individual* can affect and be affected. As such, Deleuze does not homogenize experience under the 'collective' capacities Tolia-Kelly suggests as this is reductive of the singularity of each specific manifestation of those collective categories. Deleuze then pays attention to a more radical difference; to a difference in itself which is not a difference between specific manifestations or collectives, but rather a singularity in itself.

The second debate I will now discuss has arisen regarding the argued prominence of the auto-affective and overly agentive visions of the body and embodied experience that have emerged in recent work in non-representational theory, and the resultant call for there to be more attention paid to the vulnerability and passivity of the body. As such, Harrison (2008: 423) is critical of the recent prominence of what he calls "the body in action" in recent accounts of embodied experience. Harrison (2008: 423) calls attention to the prominence of the body being apprehended as "practically and constitutively engaged in the disclosure of the world and in the creation and maintenance of meaning and signification". While not wanting to underplay the significance and positive nature of such contributions, Harrison calls attention to the potential lack of consideration being given to the ways in which the body is susceptible and passive, and wants to think about embodiment in and through its vulnerability, and, more specifically, to challenge the predominant notion of such vulnerability as something which is negative and to be overcome. Instead Harrison (2008: 427) wants to think vulnerability in terms of describing "the inherent and continuous susceptibility of corporeal life to the unchosen and the unforeseen – its inherent openness to what exceeds its abilities to contain and absorb".

Following Harrison, it is important to emphasize that affects are as much about slowing down as they are about speeding up. This is to say that there will always be a *variance* in any capacity to act. That said, while Harrison calls attention to what he deems to be the "remorseless pressure of immanence and the proliferation of becoming" in the writing of Deleuze, *there are* affects that slow us down as well as affects that speed us up in his work. There is joy *and* there is sadness. The coherence of bodies is threatened as well as potentially being expanded upon. There are 'decompositions' as much as 'compositions'. This is perhaps something that the geographical literature has not emphasized enough (so far...). This is to suggest that there is vulnerability *included* within this, but that it is not given a central position – it is not *the* aspect of our embodied existence, but *an* aspect. There is always a relative movement – a becoming-faster and a becoming-slower. As Thrift states

> *"not everything is focused intensity. Embodiment* includes *tripping, falling over, and a whole host of other such mistakes. It* includes *vulnerability, passivity, suffering, even simple hunger. It* includes *episodes of insomnia, weariness and exhaustion, a sense of insignificance and even sheer indifference to the world. In other words, bodies can and do become overwhelmed".*

While there is an *ethics* of speeding up and a call for what Deleuze calls a "bliss of action" (Deleuze 1988: 28), this is an *ethical* imperative and not a more general statement about corporeal existence which is about composition and decomposition, joy and sadness.

Social Justice

Social justice is "justice in terms of the distribution of wealth, opportunities, and privileges within a society". Classically, "justice" (especially corrective justice or distributive justice) referred to ensuring that individuals both fulfilled their societal roles, and received what was due from society. "Social justice" is generally used to refer to a set of institutions which will enable people to lead a fulfilling life and be active contributors to their community. The goal of social justice is generally the same as human development. The relevant institutions can include education, health care, social security, labour rights, as well as a broader system of public services, progressive taxation and regulation of markets, to ensure fair distribution of wealth, equality of opportunity, and no gross inequality of outcome.

While the concept of social justice can be traced through Ancient and Renaissance philosophy, such as Socrates, Thomas Aquinas, Spinoza and Thomas Paine, the term "social justice" only became used explicitly from the 1840s. A Jesuit priest named Luigi Taparelli is typically credited with coining the term, and it spread during the revolutions of 1848 with the work of Antonio Rosmini-Serbati. In the late industrial revolution, progressive American legal scholars began to use the term more, particularly Louis Brandeis and Roscoe Pound. From the early 20th century it was also embedded in international law and institutions, starting with the Treaty of Versailles 1919. The preamble to establish the International Labour Organisation recalled that "universal and lasting peace can be established only if it is based upon social justice." In the later 20th century, social justice was made central to the philosophy of the social contract, primarily by John Rawls in *A Theory of Justice* (1971). In 1993, the Vienna Declaration and Programme of Action treats social justice as a purpose of the human rights education.

History

The different concepts of justice, as discussed in ancient Western philosophy, were typically centred upon the community. Plato wrote in *The Republic* that it would be an ideal state that "every member of the community must be assigned to the class for which he finds himself best fitted." Aristotle believed rights existed only between free people, and the law should take "account in the first instance of relations of inequality in which individuals are treated in proportion to their worth and only secondarily of relations of equality." The Letter to the Ephesians attributed to Paul states that everyone should be bound to do his duty in the class where they were born. Reflecting this time when slavery and subjugation of women was typical, ancient views of justice tended to reflect the rigid class systems that still prevailed. On the other hand, for the privileged groups, strong concepts of fairness and the community existed. Distributive justice was said by Aristotle to require that people were distributed goods and assets according to their merit. Socrates (through Plato's dialogue *Crito*) is attributed developing the idea of a social contract, whereby people ought to follow the rules of a society, and accept its burdens, because they have lived to accept its benefits. During the Middle Ages, religious scholars particularly, such as Thomas Aquinas continued discussion of justice in various ways, but ultimately connected to being a good citizen for the purpose of serving God.

After the Renaissance and Reformation, the modern concept of social justice, as developing human potential, began to emerge through the work of a series of authors. Baruch Spinoza in *On the Improvement of the Understanding* (1677) contended that the one true aim of life should be to acquire "a human character much more stable than one's own", and to achieve this "pitch of perfection... The chief good is that he should arrive, together with other individuals if possible, at the possession of the aforesaid character." During the enlightenment and responding to the French and American Revolutions, Thomas Paine similarly wrote in *The Rights of Man* (1792) society should give "genius a fair and universal chance" and so "the construction of government ought to be such as to bring forward... all that extent of capacity which never fails to appear in revolutions."

The first modern usage of the specific term "social justice" is typically attributed to Catholic thinkers from the 1840s, including to the Jesuit Luigi Taparelli in *Civiltà Cattolica,* based on the work of St. Thomas Aquinas. He argued that rival capitalist and socialist theories, based on subjective Cartesian thinking, undermined the unity of society present in Thomistic metaphysics as neither were sufficiently concerned with moral philosophy. Writing in 1861, the influential British philosopher, politician and economist, John Stuart Mill stated in *Utilitarianism* his view that "Society should treat all equally well who have deserved equally well of it, that is, who have deserved equally well absolutely. This is the highest abstract standard of social and distributive justice; towards which all institutions, and the efforts of all virtuous citizens, should be made in the utmost degree to converge."

In the later 19th and early 20th century, social justice became an important theme in American political and legal philosophy, particularly with John Dewey, Roscoe Pound and Louis Brandeis. One of the prime concerns was the *Lochner era* decisions of the US Supreme Court to strike down legislation passed by state governments and the Federal government for social and economic improvement, such as the eight hour day or the right to join a trade union. After the First World War, the founding document of the International Labour Organisation took up the same terminology in its preamble, stating that "peace can be established only if it is based on social justice". From this point, discussion of social justice entered into mainstream legal and academic discourse. In the late 20th century, a number of liberal and conservative thinkers, notably Friedrich von Hayek rejected the concept by stating that it did not mean anything, or meant too many things. However the concept remained highly influential, particularly with its promotion by philosophers such as John Rawls.

Contemporary Theory

Judaism: In *To Heal a Fractured World: The Ethics of Responsibility,* Rabbi Jonathan Sacks states that social justice has a central place in Judaism. One of Judaism's most distinctive and challenging ideas is its ethics of responsibility reflected in the concepts of simcha ("gladness" or "joy"), tzedakah ("the religious obligation to perform charity and philanthropic acts"), chesed ("deeds of kindness"), and tikkun olam ("repairing the world").

Catholicism

Catholic social teaching consists of those aspects of Roman Catholic doctrine which relate to matters dealing with the respect of the individual human life. A distinctive feature of the Catholic social doctrine is their concern for the poorest and most vulnerable members of society. Two of the seven key areas of "Catholic social teaching" are pertinent to social justice:

- Life and dignity of the human person: The foundational principle of all "Catholic Social Teaching" is the sanctity of all human life and the inherent dignity of every human person, from conception to natural death. Human life must be valued above all material possessions.
- Preferential option for the poor and vulnerable: Catholics believe Jesus taught that on the Day of Judgement God will ask what each person did to help the poor and needy: "Amen, I say to you, whatever you did for one of these least brothers of mine, you did for me." The Catholic Church believes that through words, prayers and deeds one must show solidarity with, and compassion for, the poor. The moral test of any society is "how it treats its most vulnerable members. The poor have the most urgent moral claim on the conscience of the nation. People are called to look at public policy decisions in terms of how they affect the poor."

Even before it was propounded in the Catholic social doctrine, social justice appeared regularly in the history of the Catholic Church:

- Pope Leo XIII, who studied under Taparelli, published in 1891 the encyclical *Rerum Novarum* (On the Condition of the Working Classes), rejecting both socialism and capitalism, while defending labour unions and private property. He stated that society should be based on cooperation and not class conflict and competition. In this document, Leo set out the Catholic Church's response to the social instability and labour conflict that had arisen in the wake of industrialization and had led to the rise of socialism. The Pope advocated that the role of the State was to promote social justice through the protection of rights, while the Church must speak out on social issues in order to teach correct social principles and ensure class harmony.

- The encyclical *Quadragesimo Anno* (On Reconstruction of the Social Order, literally "in the fortieth year") of 1931 by Pope Pius XI, encourages a living wage, subsidiarity, and advocates that social justice is a personal virtue as well as an attribute of the social order, saying that society can be just only if individuals and institutions are just.
- Pope John Paul II added much to the corpus of the Catholic social teaching, penning three encyclicals which would deal with issues such as economics, politics, geo-political situations, ownership of the means of production, private property and the "social mortgage", and private property. The encyclicals of Laborem Exercens, Solicitudo Rei Socialis, and Centesimus Annus are just a small portion of his overall contribution to Catholic social justice. Pope John Paul II was a strong advocate of justice and human rights, and spoke forcefully for the poor. He addresses issues such as the problems that technology can present should it be misused, and admits a fear that the "progress" of the world is not true progress at all, if it should denigrate the value of the human person.
- Pope Benedict XVI's encyclical *Deus Caritas Est* ("God is Love") of 2006 claims that justice is the defining concern of the state and the central concern of politics, and not of the church, which has charity as its central social concern. It said that the laity has the specific responsibility of pursuing social justice in civil society and that the church's active role in social justice should be to inform the debate, using reason and natural law, and also by providing moral and spiritual formation for those involved in politics.
- The official Catholic doctrine on social justice can be found in the book *Compendium of the Social Doctrine of the Church*, published in 2004 and updated in 2006, by the Pontifical Council *Iustitia et Pax*.

The Catechism (§1928–1948) contain more detail of the Church's view of social justice.

Methodism

From its founding, Methodism was a Christian social justice movement. Under John Wesley's direction, Methodists became leaders in many social justice issues of the day, including the prison reform and abolitionism movements. Wesley himself was among the first to preach for slaves rights attracting significant opposition.

Today, social justice plays a major role in the United Methodist Church. The *Book of Discipline of the United Methodist Church* says, "We hold governments responsible for the protection of the rights of the people to free and fair elections and to the freedoms of speech, religion, assembly, communications media, and petition for redress of grievances without fear of reprisal; to the right to privacy; and to the guarantee of the rights to adequate food, clothing, shelter, education, and health care." The United Methodist Church also teaches Population control as part of its doctrine.

Hinduism

Ancient Hindu society was based on equality of all beings. However, to divide labour society divided itself into hundreds of tribes[Jati]. India was governed by people of non-Hindu faiths from the 8th century which caused ruptures in societal fabric. Caste is a word from the Portuguese word "casta" and caste came to define the jatis only 500 years ago. Considerable social engineering occurred during the British rule which impacted the society's self governance. There was some social injustice in which some jatis considered themselves superior to others. The present day jati hierarchy is undergoing changes for variety of reasons including 'social justice', which is a politically popular stance in democratic India. Institutionalized affirmative action has swung the pendulum. The disparity and wide inequalities in social behaviour to some of the jatis led to various reform movements in hinduism for centuries. While legally outlawed, the caste system remains strong in practice.

Islam

The Quran contains numerous references to elements of social justice. For example, one of Islam's Five Pillars is Zakât, or alms-giving. Charity and assistance to the poor – concepts central to social justice – are and have historically been important parts of the Islamic faith.

In Muslim history, Islamic governance has often been associated with social justice. Establishment of social justice was one of the motivating factors of the Abbasid revolt against the Umayyads. The Shi'a believe that the return of the *Mahdi* will herald in "the messianic age of justice" and the Mahdi along with the Messiah (Jesus) will end plunder, torture, oppression and discrimination.

For the Muslim Brotherhood the implementation of social justice would require the rejection of consumerism and communism. The Brotherhood strongly affirmed the right to private property as well as differences in personal wealth due to factors such as hard work.

However, the Brotherhood held Muslims had an obligation to assist those Muslims in need. It held that *zakat* (alms-giving) was not voluntary charity, but rather the poor had the right to assistance from the more fortunate. Most Islamic governments therefore enforce the *zakat* through taxes.

Though monetary donations are the most practiced way of zakat, Islam is deeply rooted in the tenets of volunteerism and social activism. Areas of one's communities which require assistance and beneficiaries must be a Muslim's foci if need be, rather than strictly her or his personal or superficial wants. For example, the ecological well-being of the planet (i.e.: animal rights, global warming, natural resources degradation); locally, nationally, globally, is a campaign to which every Muslim must adhere. Many Muslims practice this today by ensuring that they produce minimal waste, give to charity what they no longer need, and spend time in prayer and meditation upon the bounties of nature so as to more mindfully approach all that is provided by nature, and ultimately, Allah. Other areas of society in need may be the safety and security of minority populations, i.e.: women or persons of color, children, the elderly, the developmentally or physically disabled, animals, et al. Social justice in Islam is a tenet to which every Muslim must corroborate in his or her daily life, and without which would create a void in all their efforts towards attaining true spirituality and a connection with God. Islam lays great stress on doctorine "justice for all."

John Rawls

Political philosopher John Rawls draws on the utilitarian insights of Bentham and Mill, the social contract ideas of John Locke, and the categorical imperative ideas of Kant. His first statement of principle was made in *A Theory of Justice* where he proposed that, "Each person possesses an inviolability founded on justice that even the welfare of society as a whole cannot override. For this reason justice denies that the loss of freedom for some is made right by a greater good shared by others." A deontological proposition that echoes Kant in framing the moral good of justice in absolutist terms. His views are definitively restated in *Political Liberalism* where society is seen "as a fair system of co-operation over time, from one generation to the next.".

All societies have a basic structure of social, economic, and political institutions, both formal and informal. In testing how well these elements fit and work together, Rawls based a key test of legitimacy on the theories of social contract. To determine whether any particular system of collectively enforced social arrangements is legitimate, he argued that one must look for agreement by the people who are subject to it, but not necessarily to an objective notion of justice based on coherent ideological grounding. Obviously, not every citizen can be asked to participate in a poll to determine his or her consent to every proposal in which some degree of coercion is involved, so one has to assume that all citizens are reasonable. Rawls constructed an argument for a two-stage process to determine a citizen's hypothetical agreement:

- The citizen agrees to be represented by X for certain purposes, and, to that extent, X holds these powers as a trustee for the citizen.
- X agrees that enforcement in a particular social context is legitimate. The citizen, therefore, is bound by this decision because it is the function of the trustee to represent the citizen in this way.

This applies to one person who represents a small group (e.g., the organiser of a social event setting a dress code) as equally as it does to national governments, which are ultimate trustees, holding representative powers for the benefit of all citizens within their territorial boundaries. Governments that fail to provide for welfare of their citizens according to the principles of justice are not legitimate. To emphasise the general principle that justice should rise from the people and not be dictated by the law-making powers of governments, Rawls asserted that, "There is ... a general presumption against imposing legal and other restrictions on conduct without sufficient reason. But this presumption creates no special priority for any particular liberty." This is support for an unranked set of liberties that reasonable citizens in all states should respect and uphold — to some extent, the list proposed by Rawls matches the normative human rights that have international recognition and direct enforcement in some nation states where the citizens need encouragement to act in a way that fixes a greater degree of equality of outcome. According to Rawls, the basic liberties that every good society should guarantee are,

- Freedom of thought;
- Liberty of conscience as it affects social relationships on the grounds of religion, philosophy, and morality;
- Political liberties (e.g. representative democratic institutions, freedom of speech and the press, and freedom of assembly);
- Freedom of association;
- Freedoms necessary for the liberty and integrity of the person (viz: freedom from slavery, freedom of movement and a reasonable degree of freedom to choose one's occupation); and
- Rights and liberties covered by the rule of law.

Criticism

The concept of social justice has come under criticism from a variety of perspectives.

Many authors criticize the idea that there exists an objective standard of social justice. Moral relativists deny that there is any kind of objective standard for justice in general. Non-cognitivists, moral skeptics, moral nihilists, and most logical positivists deny the epistemic possibility of objective notions of justice. Political realists (such as Niccolò Machiavelli) believe that any ideal of social justice is ultimately a mere justification for the status quo.

Many other people accept some of the basic principles of social justice, such as the idea that all human beings have a basic level of value, but disagree with the elaborate conclusions that may or may not follow from this. One example is the statement by H. G. Wells that all people are "equally entitled to the respect of their fellowmen."

On the other hand, some scholars reject the very idea of social justice as meaningless, religious, self-contradictory, and ideological, believing that to realise any degree of social justice is unfeasible, and that the attempt to do so must destroy all liberty. Perhaps the most complete rejection of the concept of social justice comes from Friedrich Hayek of the Austrian School of economics:

There can be no test by which we can discover what is 'socially unjust' because there is no subject by which such an injustice can be committed, and there are no rules of individual conduct the observance of which in the market order would secure to the individuals and groups the position which as such (as distinguished from the procedure by which it is determined) would appear just to us. Social justice does not belong to the category of error but to that of nonsense, like the term `a moral stone'.

Ben O'Neill of the University of New South Wales argues that, for proponents of "social justice": the notion of "rights" is a mere term of entitlement, indicative of a claim for any possible desirable good, no matter how important or trivial, abstract or tangible, recent or ancient. It is merely an assertion of desire, and a declaration of intention to use the language of rights to acquire said desire.

In fact, since the program of social justice inevitably involves claims for government provision of goods, paid for through the efforts of others, the term actually refers to an intention to use *force* to acquire one's desires. Not to earn desirable goods by rational thought and action, production and voluntary exchange, but to go in there and forcibly take goods from those who can supply them!

Janusz Korwin-Mikke argues simply: "Either 'social justice' has the same meaning as 'justice' – or not. If so – why use the additional word 'social?' We lose time, we destroy trees to obtain paper necessary to print this word. If not, if 'social justice' means something different from 'justice' – then 'something different from justice' is by definition 'injustice.'"

Sociologist Carl L. Bankston has argued that a secular, leftist view of social justice entails viewing the redistribution of goods and resources as based on the rights of disadvantaged categories of people, rather than on compassion or national interest. Bankston maintains that this secular version of social justice became widely accepted due to the rise of demand-side economics and to the moral influence of the civil rights movement.

Cosmic Values

Hunter Lewis' work promoting natural healthcare and sustainable economies advocates for conservation as a key premise in social justice. His manifesto on sustainability ties the continued thriving of human life to real conditions, the environment supporting that life, and associates injustice with the detrimental effects of unintended consequences of human actions. Quoting classical Greek thinkers like Epicurus on the good of pursuing happiness, Hunter also cites ornithologist, naturalist, and philosopher Alexander Skutch in his book Moral Foundations:

> *The common feature which unites the activities most consistently forbidden by the moral codes of civilized peoples is that by their very nature they cannot be both habitual and enduring, because they tend to destroy the conditions which make them possible.*

Pope Benedict XVI cites Teilhard de Chardin in a vision of the cosmos as a 'living host' embracing an understanding of ecology that includes mankinds's relationship to fellow men, that pollution affects not just the natural world but interpersonal relations also. Cosmic harmony, justice and peace are closely interrelated:

If you want to cultivate peace, protect creation.

United Nations

The United Nations' 2006 document *Social Justice in an Open World: The Role of the United Nations*, states that "Social justice may be broadly understood as the fair and compassionate distribution of the fruits of economic growth..."

The term "social justice" was seen by the U.N. "as a substitute for the protection of human rights [and] first appeared in United Nations texts during the second half of the 1960s. At the initiative of the Soviet Union, and with the support of developing countries, the term was used in the Declaration on Social Progress and Development, adopted in 1969."

The same document reports, "From the comprehensive global perspective shaped by the United Nations Charter and the Universal Declaration of Human Rights, neglect of the pursuit of social justice in all its dimensions translates into de facto acceptance of a future marred by violence, repression and chaos." The report concludes, "Social justice is not possible without strong and coherent redistributive policies conceived and implemented by public agencies."

The same UN document offers a concise history: "The notion of social justice is relatively new. None of history's great philosophers—not Plato or Aristotle, or Confucius or Averroes, or even Rousseau or Kant—saw the need to consider justice or the redress of injustices from a social perspective. The concept first surfaced in Western thought and political language in the wake of the industrial revolution and the parallel development of the socialist doctrine. It emerged as an expression of protest against what was perceived as the capitalist exploitation of labour and as a focal point for the development of measures to improve the human condition. It was born as a revolutionary slogan embodying the ideals of progress and fraternity. Following the revolutions that shook Europe in the mid-1800s, social justice became a rallying cry for progressive thinkers and political activists.... By the mid-twentieth century, the concept of social justice had become central to the ideologies and programmes of virtually all the leftist and centrist political parties around the world..."

Social Justice Movements

Social justice is also a concept that is used to describe the movement towards a socially just world, e.g., the Global Justice Movement. In this context, social justice is based on the concepts of human rights and equality, and can be defined as *"the way in which human rights are manifested in the everyday lives of people at every level of society"*.

A number of movements are working to achieve social justice in society. These movements are working towards the realisation of a world where all members of a society, regardless of background or procedural justice, have basic human rights and equal access to the benefits of their society.

Liberation Theology

Liberation theology is a movement in Christian theology which conveys the teachings of Jesus Christ in terms of a liberation from unjust economic, political, or social conditions. It has been described by proponents as "an interpretation of Christian faith through the poor's suffering, their struggle and hope, and a critique of society and the Catholic faith and Christianity through the eyes of the poor", and by detractors as Christianity perverted by Marxism and Communism.

Although liberation theology has grown into an international and inter-denominational movement, it began as a movement within the Catholic Church in Latin America in the 1950s–1960s. It arose principally as a moral reaction to the poverty caused by social injustice in that region. It achieved prominence in the 1970s and 1980s. The term was coined by the Peruvian priest, Gustavo Gutiérrez, who wrote one of the movement's most famous books, *A Theology of Liberation* (1971). According to Sarah Kleeb, "Marx would surely take issue," she writes, "with the appropriation of his works in a religious context...there is no way to reconcile Marx's views of religion with those of Gutierrez, they are simply incompatible. Despite this, in terms of their understanding of the necessity of a just and righteous world, and the nearly inevitable obstructions along such a path, the two have much in common; and, particularly in the first edition of [A Theology of Liberation], the use of Marxian theory is quite evident."

Other noted exponents are Leonardo Boff of Brazil, Jon Sobrino of El Salvador, and Juan Luis Segundo of Uruguay.

Health Care

Social justice has more recently made its way into the field of bioethics. Discussion involves topics such as affordable access to health care, especially for low income households and families. The discussion also raises questions such as whether society should bear healthcare costs for low income families, and whether the global marketplace is a good thing to deal with healthcare. Ruth Faden and Madison Powers of the Johns Hopkins Berman Institute of Bioethics focus their analysis of social justice on which inequalities matter the most. They develop a social justice theory that answers some of these questions in concrete settings.

Social injustices occur when there is a preventable difference in health states among a population of people. These social injustices take on the form of health inequities when negative health states such as malnourishment, and infectious diseases are more prevalent among an impoverished nation. These negative health states can often be prevented by providing social and economic structures such as Primary Healthcare which ensure the general population has equal access to health care services regardless of income level, gender, education or any other stratifying factor. Integrating social justice to health inherently reflects the social determinants of health model without discounting the role of the bio-medical model.

Human Rights Education

The Vienna Declaration and Programme of Action affirm that "Human rights education should include peace, democracy, development and social justice, as set forth in international and regional human rights instruments, in order to achieve common understanding and awareness with a view to strengthening universal commitment to human rights."

Sexuality and Space

Sexuality and space is a field of study within human geography. The phrase encompasses all relationships and interactions between human sexuality, space and place, themes studied within, but not limited to cultural geography, i.e. environmental and architectural psychology, urban sociology, gender studies, queer studies, socio-legal studies, planning, housing studies and criminology.

Specific topics which fall into this area are the geographies of LGBT residence, public sex environments, sites of queer resistance, global sexualities, sex tourism, the geographies of prostitution and adult entertainment, use of sexualised locations in the arts, and sexual citizenship. The field is now well represented within academic curricula at University level, and is beginning to make its influence felt on secondary level education (in the UK).

Origins and Development

The work of sociologists has long been concerned with the relationship between urbanization and sexuality, especially in the form of visible clusters or neighbourhoods typified by specific sexual moralities or practices. Identification of 'vice areas' and, latterly, 'gay villages', has been a stock in trade of urban sociology since at least the time of the Chicago School.

The origins of the term "Sexuality and Space" can be traced back to the early 1990s where usage of the phrase was popularized by two publications. In 1990 what may be described as 'Gay Geography' was presented to a wider audience when an article by Larry Knopp was published in the *Geographical Magazine* to some controversy. In 1992 Beatriz Colomina's *Sexuality and Space (Princeton Papers on Architecture)* was released; in which the term is used to elaborate on the symbolism of towers and other structures as Phallic icons. The paper goes on to discuss the sexual psychology of colour and other design elements. A review of the papers was released by Elizabeth Wilson in Harvard Design Magazine, Winter/Spring 1997.

Within contemporary geography, studies of sexuality are primarily social and cultural in orientation, though there is also notable engagement with political and economic geography, particular in work on the rise of queer autonomous spaces, economies and alternative (queer) capitalisms. Much work is informed by a politics intended to oppose homophobia and heterosexism, inform sexual health, and promote more inclusive forms of sexual citizenship. Methodologically, much work has been qualitative in orientation, rejecting traditional 'straight' methodologies, yet quantitative methods and GIS have also be utilised to good effect. Work remains predominantly focused on the metropolitan centres of the urban West, but there have been notable studies that focus on rural sexualities and sexualities in the global South.

Geographies of LGBT

Although 'minority' sexuality remains a topic that hardly gets a mention in school geography, it has become an accepted part of many university geography departments and is often taught as part of courses on Social and Cultural Geography. Arguably, the most influential book-publication to position sexuality as an accepted part of geography was *Mapping Desire*, an edited collection by David Bell and Gill Valentine. Bell and Valentine provide a critical review of the history of geographical works on sexuality and set an agenda for further research. They are especially critical of the earliest sexual geographies written during the 1970s and 1980s in the UK and North America. In contrast to the 'dots on maps' approach of the 1970s and 1980s, *Mapping Desire* represents an attempt to map the geographies of homosexuality, trans-sexuality, bisexuality, sadomasochism and butch-femme lesbian identities. This represented an important landmark in geographer's engagement with, and development of, queer theory. Subsequent research has developed this work, with an increasing focus on transnational LGBT activism; the intersections of nationhood and sexuality and questions of LGBT citizenship and sexual politics at scales from the body to the global.

Ever since the rise of attention of geographies of LGBT in the last 1970s and 1980s, more researches had been focused on the relationship between place, space and LGBT group afterward. New phenomenons and issues are being explore, for example, in Dereka Rushbrook research, he points out that some of the secondary cities in US like

Portland, Oregon or Austin, Texas had seen the Gay village in their cities as something that represent the modernness and diversity of their cites; furthermore, transsexual and third gender, so called the Kathoey in Thailand are worldwide famous for their dance performances and it considers as something must watch while visiting Thailand.

On the other side, people had started to realise that the potential consumption of the LGBT group and cities are starting to targeting the group which creates the phenomenon called pink tourism or LGBT tourism where providing safe yet non-discriminate service and facilities like pub and sauna that targeting the LGBT group. Even more, the increasing legalize of the same sex marriage in some of the western countries had significant impact of changing the LGBT groups migrating and travelling pattern. Although, the marriage might not has the same legal power back in their countries.

As Robert Aldrichsaid in the article, Homosexuality and the City :An Historical Overview, there is an inseparable relationship between land and people where people are constantly shaping the landscape .For example, one of the school in Thailand offered to build a third gender washroom in order to reduce uncomfortable feeling of transsexual students using either male or female washroom. And on same hand, the relationship of sexuality and space is not independent from other areas of geography; often, there are other aspects associate with those issues and phenomenons like cultural geography and political geography. In 2012, further concern has been made toward the LGBT community and retirement home in British Columbia, Canada. The Vancouver man, "Alex Sangha is determined to make sure elderly LGBT people have a comfortable place to spend their twilight years." By building retirement home for LGBT community where Montreal, Canada has already has one operating as well as a few in the United States.

Nowadays, with the rising awareness of LGBT communities around the world, LGBT influences would play a more important role in shaping our space with no doubt.

Heterosexual Geographies

Research on Sexualities and Space has widened over time to encompass studies not just of LGBT populations but also the geographies and spaces of heterosexualities. This has included, inter alia, consideration of the impacts of sexuality on the visibilities of commercial sex; the design and consumption of housing; spaces of sex education; the sexualisation of leisure and retail spaces; landscapes of sex tourism; spaces of love, caring and intimacy. This has bought geographies of sexuality into dialogue with gender geography by showing that sexual norms reproduce particular ideas of masculinity and femininity.

Spiritual Spaces of LGBTQI

Members of non-heterosexual communities may lose the ability to worship or practice religion within the most widespread religious organisations due to the fact that they are not accepted as a consequence of their sexuality. Therefore other spiritual spaces are created, known in many cases as "Queer Spiritual Spaces", that can vary from sacred buildings or locations that can be "queered" to natural environments or cultural practices in themselves. This kind of behaviour leads to the general population believing that LGBTQI (Lesbian, Gay, Bisexual, Transgendered, Queer and Intersex) communities are largely atheist or agnostic, but instead some are just adopting non-Abrahamic faiths with pre-Christian traditions and customs, that are based on aspirational encouragement and personal well-being.

The other option to creating these "Queer Spiritual Spaces" is taking part in the "queering" of religions, a movement which allows people to still practice their traditional religions but within a space that is tolerant of their sexual choices. This forms a new way of cultural engagement and opens up religious groups to people who would otherwise be outcasts. This option is becoming more and more popular and therefore queer churches have become an alternative for the LGBTQI communities around the world.

Alternate religions/spiritual organisations have been created as well, such as the Lesbian stem of Separatist feminism that managed to give a voice to a portion of the community that was ostracized, and also had the power to give a foundation or reasoning for the differences they encountered with heterosexual society.

Geographies of Sex Commerce

Research on the location of vice and prostitution have long been associated with the study of sexuality and space. Pioneering – if controversial – in this area was Symanski's (1988) *Immoral Landscape*; subsequent studies have considered the socio-legal regulation of spaces of prostitution, adult entertainment, sex shops and hostess bars, and sought to place such issues in a wider theoretical context relating to the reproduction of heteronormality. Much work, however, ignores male sex work as well as forms of intersex and transwork, whilst other work continues to focus solely on the relationship between locations of sex working and the distribution of sexual infection, including HIV. Studies such as *Risk Assessment of Long-Haul Truck Drivers* by the University of Alabama at Birmingham, active since April 2007, may also be related to this field of study as the statistics gathered will represent sampling of sexual behaviour in a controlled population of a subgroup.

Criticisms & Conflicts

There have been several critiques of the field, as well as conflicts within the discipline. Studies of sexuality and space has been criticized for universalizing a Western-centric position that has minimal relevance beyond the urbanized Western world. These ideas of sexuality constitute a new homonormativity, which typically privileges white, middle-class males, to the exclusion of trans people, the lower class, and people of colour. Institutions meant to create non-heteronormative spaces, such as the Gay Games, are only accessible to those who are able to afford registration fees, airfare and training, and remain predominantly white. Gender differences are also erased in adopting a "queer" identity. Some assert that by foregoing the gendered term "lesbian" for "queer", women are left unrecognised by such a universalizing signifier - like when women are incorporated under "mankind". Thus, while progress is made on challenging heteronormativity, queer studies have been criticized for potentially reinforcing other forms of marginalization.

Early work on lesbian and gay geographies throughout the 1990s was done by academics working in American universities, and focused almost exclusively on the lives of those in the global North. This is problematic as the queer identity is sometimes used as a global, all-encompassing identity for the LGBT community, thus imposing Western notions of sexuality on all other cultures. Such ideas include unchallenged assumptions as to the nature of "gay rights", and what proper liberation looks like. As a result, certain cultures are labelled "forward" or "backward" based on a Western conception of the queer identity, and the cultural nuances and diversity of other sexualities are left unrecognised.

Conflicts within studies of sexuality and space also exist. One such conflict is between "assimilationist" and "liberationist" perspectives on LGBT spaces. Toronto's gay village is a site of such conflict. Assimilationists are against the creation of a "gay ghetto" in Toronto, advocating instead for the integration of LGBT people into suburbs, to show that they are just like everybody else. Liberationists see the gay village as too commercial to develop the radical, activist community they see as necessary for LGBT rights. As such, Toronto's gay village is not simply a homogenous space for sexual dissidents, but is an unfixed and contested space.

Organisations

Within the discipline of Geography, initial scepticism and even opposition to research on sexuality has given way to recognition that geographies of sexuality offer an important perspective on the relationship between people and place (albeit that some continue to regard the area as of marginal importance).

Phobia

In clinical psychology, a phobia is a type of anxiety disorder, usually defined as a persistent fear of an object or situation in which the sufferer commits to great lengths in avoiding, typically disproportional to the actual danger posed, often being recognised as irrational. In the event the phobia cannot be avoided entirely, the sufferer will endure the situation or object with marked distress and significant interference in social or occupational activities.

The terms *distress* and *impairment* as defined by the *Diagnostic and Statistical Manual of Mental Disorders, Fourth Edition* (DSM-IV-TR) should also take into account the context of the sufferer's environment if attempting a diagnosis. The DSM-IV-TR states that if a phobic stimulus, whether it be an object or a social situation, is absent entirely in an environment — a diagnosis cannot be made. An example of this situation would be an individual who has a fear of mice (Suriphobia) but lives in an area devoid of mice.

Even though the concept of mice causes marked distress and impairment within the individual, because the individual does not encounter mice in the environment no actual distress or impairment is ever experienced. Proximity and the degree to which escape from the phobic stimulus is impossible should also be considered. As the sufferer approaches a phobic stimulus, anxiety levels increase (e.g. as one gets closer to a snake, fear increases in ophidiophobia), and the degree to which escape of the phobic stimulus is limited has the effect of varying the intensity of fear in instances such as riding an elevator (e.g. anxiety increases at the midway point between floors and decreases when the floor is reached and the doors open).

The term *phobia* is encompassing and usually discussed in the contexts of specific phobias and social phobias. Specific phobias are nouns, such as *arachnophobia* or *acrophobia*, which are specific, and social phobias are phobias within social situations, such as public speaking and crowded areas. Some phobias, such as xenophobia, overlap with many other phobias.

Classification

Clinical: Most phobias are classified into three categories and, according to the *Diagnostic and Statistical Manual of Mental Disorders, Fourth Edition* (DSM-IV), such phobias are considered to be sub-types of anxiety disorder. The three categories are:

1. Social phobia: fear of other people or social situations such as performance anxiety or fears of embarrassment by scrutiny of others. Overcoming social phobia is often very difficult without the help of therapy or support groups. Social phobia may be further subdivided into
   - generalized social phobia (also known as social anxiety disorder or simply social anxiety).
   - specific social phobia, in which anxiety is triggered only in specific situations. The symptoms may extend to psychosomatic manifestation of physical problems. For example, sufferers of paruresis find it difficult or impossible to urinate in reduced levels of privacy. This goes far beyond mere preference: when the condition triggers, the person physically cannot empty their bladder.
2. Specific phobias: fear of a double specific panic trigger such as spiders, snakes, dogs, wind, water, heights, flying, catching a specific illness, etc. Many people have these fears but to a lesser degree than those who suffer from specific phobias. People with the phobias specifically avoid the entity they fear.
3. Agoraphobia: a generalized fear of leaving home or a small familiar 'safe' area, and of possible panic attacks that might follow. It may also be caused by various specific phobias such as fear of open spaces, social embarrassment (social agoraphobia), fear of contamination (fear of germs, possibly complicated by obsessive-compulsive disorder) or PTSD (post traumatic stress disorder) related to a trauma that occurred out of doors.

Phobias vary in severity among individuals. Some individuals can simply avoid the subject of their fear and suffer relatively mild anxiety over that fear. Others suffer full-fledged panic attacks with all the associated disabling symptoms. Most individuals understand that they are suffering from an irrational fear, but are powerless to override their panic reaction.

Specific Phobias

A specific phobia is a marked and persistent fear of an object or situation which brings about an excessive or unreasonable fear when in the presence of, or anticipating, a specific object; the specific phobias may also include concerns with losing control, panicking, and fainting which is the direct result of an encounter with the phobia. Specific phobias are defined in relation to objects or situations whereas social phobias emphasize social fear and the evaluations that might accompany them.

The DSM breaks specific phobias into five subtypes: animal, natural environment, blood-injection-injury, situational, and other. In children, phobias involving animals, natural environment (darkness), and blood-injection-injury usually develop between the ages of 7 and 9, and these are reflective of normal development. Additionally, specific phobias are most prevalent in children between ages 10 and 13.

Social Phobia

Unlike specific phobias, social phobias include fear of public situations and scrutiny which leads to embarrassment or humiliation in the diagnostic criteria. People with social phobia have extreme feelings of self-consciousness built into powerful fear. In social phobias, there is also a generalized category. Unlike specific phobias which may develop before the age of 10, social phobias are typically not present until pubertal transition. After this transition, the prevalence of social phobia increases with age. Many adolescents who develop a social phobia consequently become rejected by their peers. As interpersonal dysfunction is a risk factor for depression, there are some negative outcomes for adolescents with social phobia. For example, about 20% of adolescents diagnosed with a social anxiety disorder also suffer from depression and use alcohol or other substances.

Cause

Environmental: Rachman proposed three pathways to acquiring fear conditioning: classical conditioning, vicarious acquisition and informational/instructional acquisition:

- Much of the progress in understanding the acquisition of fear responses in phobias can be attributed to the Pavlovian model, which is synonymous with classical conditioning. When an aversive stimulus and a neutral one are paired together, for instance when an electric shock is given in a specific room, the subject can start to fear not only the shock but the room as well. In behavioural terms, this is described as a conditioned stimulus (CS) *(the room)* that is paired with an aversive unconditioned stimulus (UCS) *(the shock)*, which leads to a conditioned response (CR) *(fear for the room)* (CS+UCS=CR).

For instance, in case of the fear of heights (acrophobia), the CS is heights such as a balcony on the top floors of a high rise building. The UCS originates from an aversive or traumatizing event in the person's life, such as almost falling down from a great height. The original fear of almost falling down is associated with being on a high place, leading to a fear of heights. In other words, the CS *(heights)* associated with the aversive UCS *(almost falling down)* leads to the CR *(fear)*.

This direct conditioning model, though very influential in the theory of fear acquisition, is not the only way to acquire a phobia.

- Vicarious fear acquisition is learning to fear something, not by a subject's own experience of fear, but by watching others reacting fearfully (observational learning). For instance, when a child sees a parent reacting fearfully to an animal, the child can become afraid of the animal as well.
- Informational/instructional fear acquisition is learning to fear something by getting information. For instance, fearing electrical wire after having heard that touching it will result in an electric shock.

A conditioned fear response to an object or situation is not always a phobia. To meet the criteria for a phobia there must also be symptoms of impairment and avoidance. Impairment is defined as being unable to complete routine tasks whether occupational, academic or social. In acrophobia an impairment of occupation could result from not taking a job solely because of its location at the top floor of a building, or socially not participating in a social event at a theme park. The avoidance aspect is defined as behaviour that results in the omission of an aversive event that would otherwise occur with the goal of the preventing anxiety.

Mechanism

Beneath the lateral fissure in the cerebral cortex, the insula, or insular cortex, of the brain has been identified as part of the limbic system, along with cingulated gyrus, hippocampus, corpus callosum, and other nearby cortices. This system has been found to play a role in emotion processing and the insula, in particular, may contribute through its role in maintaining autonomic functions. Studies by Critchley et al. indicate the insula as being involved in the experience of emotion by detecting and interpreting threatening stimuli. Similar studies involved in monitoring the activity of the insula show a correlation between increased insular activation and anxiety.

In the frontal lobes, other cortices involved with phobia and fear are the anterior cingulate cortex and the medial prefrontal cortex. In the processing of emotional stimuli, studies on phobic reactions to facial expressions have indicated these areas to be involved in processing and responding to negative stimuli. The ventromedial prefrontal cortex has been said to influence the amygdala by monitoring its reaction to emotional stimuli or even fearful memories. Most specifically, the medial prefrontal cortex is active during extinction of fear and is responsible for long term extinction. Stimulation of this area decreases conditioned fear responses and so its role may be in inhibiting the amygdala and its reaction to fearful stimuli.

The hippocampus is a horseshoe shaped structure that plays an important part in the brain's limbic system because of its role in forming memories and connecting them with emotions and the senses. When dealing with fear, the hippocampus receives impulses from the amygdala that allows it to connect the fear with a certain sense, such as a smell or sound.

Amygdala

The amygdala is an "almond shaped" mass of nuclei that is located deep in the brain's medial temporal lobe. It processes the events associated with fear and is being linked to anxiety disorders and social phobias. The amygdala's ability to respond to fearful stimuli occurs through the process of fear conditioning. Similar to classical conditioning, the amygdala learns to associate a conditioned stimulus with a negative or avoidant stimulus, creating a conditioned fear response that is often seen in phobic individuals. In this way the amygdala is responsible for not only recognising certain stimuli or cues as dangerous, but plays a role in the storage of threatening stimuli to memory. The basolateral nuclei (or basolateral amygdala) and the hippocampus interact with the amygdala in the storage of memory, which suggests why memories are often remembered more vividly if they have emotional significance.

In addition to memory, the amygdala also triggers the secretion of hormones that affect fear and aggression. When the fear or aggression response is initiated, the amygdala releases hormones into the body to put the human body into an "alert" state, which prepares the individual to move, run, fight, etc. This defencive "alert" state and response is generally referred to in psychology as the fight-or-flight response.

Inside the brain, however, this stress response can be observed in the hypothalamic-pituitary-adrenal axis (HPA). This circuit incorporates the process of receiving stimuli, interpreting it, and releasing certain hormones into the blood stream. The parvocellular neurosecretary neurons of the hypothalamus release corticotropin-releasing hormone(CRH) which is sent to the anterior pituitary. Here the pituitary releases adrenocorticotropic hormone (ACTH) which ultimately stimulates the release of cortisol. In relation to anxiety, the amygdala is responsible for activating this circuit, while the hippocampus is responsible for suppressing it. Glucocorticoid receptors in the hippocampus monitor the amount of cortisol in the system and through negative feedback can tell the hypothalamus to stop releasing CRH.

Studies on mice engineered to have high concentrations of CRH showed higher levels of anxiety, while those engineered to have no or low amounts of CRH receptors were less anxious. In phobic patients, therefore, high amounts of cortisol may be present, or alternatively, there may be low levels of glucocorticoid receptors or even serotonin (5-HT).

Disruption by Damage

For the areas in the brain involved in emotion—most specifically fear— the processing and response to emotional stimuli can be significantly altered when one of these regions becomes lesioned or damaged. Damage to the cortical areas involved in the limbic system such as the cingulate cortex or frontal lobes have resulted in extreme changes in emotion. Other types of damage include Klüver-Bucy Syndrome and Urbach-Wiethe disease. In Klüver-Bucy syndrome, a temporal lobectomy, or removal of the temporal lobes results in changes involving fear and aggression. Specifically, the removal of these lobes results in decreased fear, confirming its role in fear recognition and response. Bilateral damage to the medial temporal lobes, which is known as Urbach-Wiethe disease exhibits similar symptoms of decreased fear and aggression, but also an inability to recognise emotional expressions, especially angry or fearful faces.

The amygdala's role in learned fear includes interactions with other brain regions in the neural circuit of fear. While lesions in the amygdala can inhibit its ability to recognise fearful stimuli, other areas such as the ventromedial prefrontal cortex and the basolateral nuclei of the amygdala can affect the region's ability to not only become conditioned to fearful stimuli, but to eventually extinguish them. The basolateral nuclei, through receiving stimulus info, undergo synaptic changes which allow the amygdala to develop a conditioned response to fearful stimuli. Lesions in this area, therefore, have been shown to disrupt the acquisition of learned responses to fear. Likewise, lesions in the ventromedial prefrontal cortex (the area responsible for monitoring the amygdala) have been shown to not only slow down the speed of extinguishing a learned fear response, but also how effective or strong the extinction is. This suggests there is a pathway or circuit among the amygdala and nearby cortical areas that process emotional stimuli and influence emotional expression, all of which can be disrupted when an area becomes damaged.

Treatments

Various methods are claimed to treat phobias. Some therapists use virtual reality or imagery exercise to desensitize patients to the feared entity. These are parts of systematic desensitization therapy.

Therapy

Cognitive behavioural therapy (CBT) can be beneficial. Cognitive behavioural therapy allows the patient to challenge dysfunctional thoughts or beliefs by being mindful of their own feelings with the aim that the patient will realise their fear is irrational. CBT may be conducted in a group setting. Gradual desensitisation treatment and CBT are often successful, provided the patient is willing to endure some discomfort. In one clinical trial, 90% of patients were observed to no longer have a phobic reaction after successful CBT treatment.

CBT is also an effective treatment for phobias in children and adolescents, and it has been adapted to be appropriate for use with this age. One example of a CBT program targeted towards children is the Coping Cat. This treatment program can be used with children between the ages of 7 and 13 to treat social phobia. This program works to decrease negative thinking, increase problem solving, and to provide a functional coping outlook in the child. Another CBT program was developed by Ann Marie Albano to treat social phobia in adolescents. This program has five stages: Psychoeducation, Skill Building, Problem Solving, Exposure, and Generalization and Maintenance. Psycho education focuses on identifying and understanding symptoms. Skill Building focuses on learning cognitive restructuring, social skills, and problem solving skills. Problem Solving focuses on identifying problems and using a proactive approach to solving them. Exposure involves exposing the adolescent to social situations in a hierarchical approach. Finally, Generalization and Maintenance involves practicing the skills learned.

Eye Movement Desensitization and Reprocessing (EMDR) has been demonstrated in peer-reviewed clinical trials to be effective in treating some phobias. Mainly used to treat Post-traumatic stress disorder, EMDR has been demonstrated as effective in easing phobia symptoms following a specific trauma, such as a fear of dogs following a dog bite.

Another method psychologists and psychiatrists use to treat patients with extreme phobias is prolonged exposure. Prolonged exposure is used in psychotherapy when the person with the phobia is exposed to the object of their fear over a long period of time. This technique is only tested when a person has overcome avoidance of or escape from the phobic object or situation. People with slight distress from their phobias usually do not need prolonged exposure to their fear.

For children and adolescents, one of the most effective treatments for specific phobias is participant modelling and reinforced practice. In this treatment method, the therapist models for the child how they should respond to their fears and then encourages the child to practice this behaviour and reinforces their efforts.

Medications

Antidepressant medications such as SSRIs or MAOIs may be helpful in some cases of phobia. Benzodiazepines may be useful in acute treatment of severe symptoms, but the risk-benefit ratio is against their long-term use in phobic disorders.

Hypnotherapy

There is little high-quality evidence looking at hypnosis. Tentative evidence suggests benefits from self-hypnosis for specific phobias.

Epidemiology

Phobias are a common form of anxiety disorders and distributions are heterogeneous by age and gender. An American study by the National Institute of Mental Health (NIMH) found that between 8.7 percent and 18.1 percent of Americans suffer from phobias, making it the most common mental illness among women in all age groups and the second most common illness among men older than 25. Between 4 percent and 10 percent of all children experience specific phobias during their lives, and social phobias occur in one percent to three percent of children and adolescents.

A Swedish study found that females have a higher incidence than males (26.5 percent for females and 12.4 percent for males). Among adults, 21.2 percent of women and 10.9 percent of men have a single specific phobia, while multiple phobias occur in 5.4 percent of females and 1.5 percent of males. Women are nearly four times as likely as men to have a fear of animals (12.1 percent in women and 3.3 percent in men) — a higher dimorphic than with all specific or generalized phobias or social phobias. Social phobias are more common in girls than in boys, while situational phobia occurs in 17.4 percent of women and 8.5 percent of men.

Society and Culture

Terminology: The word *phobia* comes from the Greek: öüâïò (*phóbos*), meaning "aversion", "fear", or "morbid fear". In popular culture, it is common for specific phobias to be given a name based on a Greek word for the object of the fear, plus the suffix *-phobia*. Creating these terms is something of a word game. Few of these terms are found in medical literature. The word *phobia* may also refer to conditions other than true phobias. For example, the term *hydrophobia* is an old name for rabies, since an aversion to water is one of that disease's symptoms. A specific phobia to water is called aquaphobia instead. A hydrophobe is a chemical compound which repels water. Similarly, the term *photophobia* usually refers to a physical complaint (aversion to light due to inflamed eyes or excessively dilated pupils), rather than an irrational fear of light.

Terms for Prejudice

A number of terms with the suffix -phobia are used non-clinically. Such terms are primarily understood as negative attitudes towards certain categories of people or other things, used in an analogy with the medical usage of the term. Usually these kinds of "phobias" are described as fear, dislike, disapproval, prejudice, hatred, discrimination, or hostility towards the object of the "phobia". Often this attitude is based on prejudices and is a particular case of most xenophobia.

Below are some examples:

- Biphobia - Negative attitudes and feelings towards bisexuality and bisexual people as a social group or as individuals.
- Homophobia - Negative attitudes and feelings toward homosexuality or people who are identified or perceived as being lesbian, gay, bisexual or transgender (LGBT).
- Islamophobia - Negative attitudes and feelings towards Islam or Muslims.
- Transphobia - Negative attitudes and feelings towards transsexualism and transsexual or transgender people, based on the expression of their internal gender identity.
- Xenophobia – fear or dislike of strangers or the unknown, sometimes used to describe nationalistic political beliefs and movements.

Spatial Justice

Spatial justice links together social justice and space. The organisation of space is a crucial dimension of human societies and reflects social facts and influences social relations (Henri Lefebvre, 1968, 1972). Consequently, both justice and injustice become visible in space. Therefore, the analysis of the interactions between space and society is necessary to understand social injustices and to formulate territorial policies aiming at tackling them. It is at this junction that the concept of spatial justice has been developed.

Spatial Justice: a Spatial Turn in the Claim for Social Justice?

Space being a fundamental dimension of human societies, social justice is embedded in it. So the understanding of interactions between space and societies is essential to the understanding of social injustices and to a reflection on planning policies that aims at reducing them. This reflection can be guided by the concept of spatial justice, which ties Social Justice with space. Spatial justice is a crucial challenge because it is the ultimate goal of many planning policies. However, the diversity of definitions of "Justice" (and of the possible "social contracts" that legitimate them), is high and the political objectives of regional planning or urban planning can be quite different and even contradictory.

Therefore, it is important to analyze the concept of spatial justice, which is still rarely questioned (particularly since the work of Anglo-American radical geographers in the 1970s–1980s to the extent that it has been taken for granted. These past few years, several events and publications have demonstrated the rising interest of human and social sciences for the concept of spatial justice.

Spatial Justice between Issues of Redistribution and Decision-Making Processes

The concept of Spatial Justice opens up several perspectives for social sciences. Building on the work of several famous Justice philosophers (John Rawls, 1971; Iris Marion Young, 1990, 2000), two contrasting approaches of justice have polarized the debate: one focuses on redistribution issues, the other concentrates on decision-making processes. A first set of approaches consists in asking questions about spatial or socio-spatial distributions and working to achieve an equal geographical distribution of society's wants and needs, such as job opportunities, access to health care, good air quality, et cetera. This is of particular concern in regions where the population has difficulty moving to a more spatially just location due to poverty, discrimination, or political restrictions (such as apartheid pass laws). Even in free, developed nations, access to many places are limited. Geographer Don Mitchell points to the mass privatization of once-public land as a common example of spatial injustice. In this distributive justice perspective, the access to material and immaterial goods, or to social positions indicates whether the situation is fair or not. At the scale of urban space, questions of accessibility, walkability and transport equity can also be seen as matters of distribution of spatial resources.

Another way of tackling the concept of spatial justice is to focus on decision-making procedures: this approach also raises issues of representations of space, of territorial or other identities and of social practices. For instance, focusing on minorities allows to explore their spatial practices but also to investigate how these are experienced and managed by various agents: this may lead to reveal forms of oppression or discrimination that a universalist approach might disregard otherwise. In sum, depending on the chosen approach, either questions are asked about spatial distributions because justice is evaluated from "results", or questions are asked about space representations, (spatial or not) identities and experiences because justice is defined as a process. Spatial justice stands as a unifying concept for the social sciences: its coherence stems from a reflection on the modalities of the political decision-making and on the policies implemented in order to improve spatial distributions.

Environmental Justice

The emergence of the concept of sustainable development has also fostered a debate on environmental equity. It questions our ontological relationship to the world, and the possibility of a fair policy addressing the needs of mankind, present and future, local and global, and of new forms of governance. The notion of "Environmental Justice" was created in the 1970s–1980s in North American cities to denounce the spatial overlapping between forms of racial discrimination and social-economic exclusion, industrial pollutions and vulnerability to natural hazards.

Chapter 6

Geopolitics

Geopolitics is the study of the effects of geography (both human and physical) on international politics and international relations. Geopolitics is a method of foreign policy analysis which seeks to understand, explain, and predict international political behaviour primarily in terms of geographical variables. Typical geographical variables are the physical location, size, climate, topography, demography, natural resources, and technological advances of the state being evaluated. Traditionally, the term has applied primarily to the impact of geography on politics, but its usage has evolved over the past century to encompass wider connotations.

Geopolitics traditionally studies the links between political power and geographic space, and examines strategic prescriptions based on the relative importance of land power and sea power in world history. The geopolitical tradition had some consistent concerns with geopolitical correlations of power in world politics, the identification of international core areas, and the relationships between naval and terrestrial capabilities. Academically, the study of geopolitics analyses geography, history, and social science with reference to spatial politics and patterns at various scales. Also, the study of geopolitics includes the study of the ensemble of relations between the interests of international political actors, interests focused to an area, space, geographical element or ways, relations which create a geopolitical system. Geopolitics is multidisciplinary in scope, and includes all aspects of the social sciences—with particular emphasis on political geography, international relations, the territorial aspects of political science and international law. The practice directly and indirectly impacts businesses and economies.

The term "Geopolitics" was coined at the beginning of the twentieth century by Rudolf Kjellén, a Swedish political scientist, who was inspired by the German geographer Friedrich Ratzel. Ratzel published *Politische Geographie (political geography)* in 1897; that book was later popularized in English by the Austro-Hungarian historian Emil Reich and the American diplomat Robert Strausz-Hupé (a faculty member of the University of Pennsylvania). Although Halford Mackinder had a pioneering role in the field, he never used the term geopolitics himself.

American Geopolitical Doctrine

Alfred Thayer Mahan and Sea Power: Alfred Thayer Mahan, a frequent commentator on world naval strategic and diplomatic affairs, believed that national greatness was inextricably associated with the sea—and particularly with its commercial use in peace and its control in war. Mahan's theoretical framework came from Antoine-Henri Jomini, and emphasized that strategic locations (such as chokepoints, canals, and coaling stations), as well as quantifiable levels of fighting power in a fleet, were conducive to control over the sea. He proposed six conditions required for a nation to have sea power:

1. Advantageous geographical position;
2. Serviceable coastlines, abundant natural resources, and favourable climate;
3. Extent of territory
4. Population large enough to defend its territory;
5. Society with an aptitude for the sea and commercial enterprise; and
6. Government with the influence and inclination to dominate the sea.

Emil Reich

The Austro-Hungarian historian Emil Reich (1854–1910) is considered to be the first having coined the acceptance in English as early as 1902 and later in 1904 in his book *Foundations of Modern Europe*.

Mackinder and the Heartland Theory

Sir Halford Mackinder's Heartland Theory initially received little attention outside geography, but would later influence the foreign policies of world powers. His formulation of the Heartland Theory was set out in his article entitled "The Geographical Pivot of History", published in England in 1904. Mackinder's doctrine of geopolitics involved concepts diametrically opposed to the notion of Alfred Thayer Mahan about the significance of navies (he coined the term *sea power*) in world conflict. He saw navy as a basis of Colombian era empire (roughly from 1492 to the nineteenth century), and predicted the twentieth century to be domain of land power. The Heartland theory hypothesized a huge empire being brought into existence in the Heartland—which wouldn't need to use coastal or transoceanic transport to remain coherent. The basic notions of Mackinder's doctrine involve considering the geography of the Earth as being divided into two sections: the World Island or Core, comprising Eurasia and Africa; and the Peripheral "islands", including the Americas, Australia, Japan, the British Isles, and Oceania. Not only was the Periphery noticeably smaller than the World Island, it necessarily required much sea transport to function at the technological level of the World Island—which contained sufficient natural resources for a developed economy.

Mackinder posited that the industrial centres of the Periphery were necessarily located in widely separated locations. The World Island could send its navy to destroy each one of them in turn, and could locate its own industries in a region further inland than the Periphery (so they would have a longer struggle reaching them, and would face a well-stocked industrial bastion). Mackinder called this region the *Heartland*. It essentially comprised Central and Eastern Europe: Ukraine, Western Russia, and Mitteleuropa. The Heartland contained the grain reserves of Ukraine, and many other natural resources. Mackinder's notion of geopolitics was summed up when he said:

> Who rules Central and Eastern Europe commands the Heartland. Who rules the Heartland commands the World-Island. Who rules the World-Island commands the World.

His doctrine was influential during the World Wars and the Cold War, as Germany and later Russia each made territorial strides toward the "Heartland".

Spykman and the "Rimland"

Nicholas J. Spykman is both a follower and critic of geostrategists Alfred Mahan, and Halford Mackinder. His work is based on assumptions similar to Mackinder's, including the unity of world politics and the world sea. He extends this to include the unity of the air. Spykman adopts Mackinder's divisions of the world, renaming some:

1. The Heartland;
2. The Rimland (analogous to Mackinder's "inner or marginal crescent" also an intermediate region, lying between the Heartland and the marginal sea powers); and

3. The Offshore Islands & Continents (Mackinder's "outer or insular crescent").

Under Spykman's theory, a Rimland separates the Heartland from ports that are usable throughout the year (that is, not frozen up during winter). Spykman suggested this required that attempts by Heartland nations (particularly Russia) to conquer ports in the Rimland must be prevented. Spykman modified Mackinder's formula on the relationship between the Heartand and the Rimland (or the inner crescent), claiming that "Who controls the rimland rules Eurasia. Who rules Eurasia controls the destinies of the world." This theory can be traced in the origins of Containment, a U.S. policy on preventing the spread of Soviet influence after World War II.

Huntington, Brzezinski and the Grand Chessboard

Since the time of Spykman, the word *geopolitics* has been applied to other theories—most notably Huntington's theory of a Clash of Civilizations and Braudel's *Grammaire des civilizations.* In a peaceable world, neither sea lanes nor surface transport are threatened, so all countries are effectively close enough to one another physically; rather, it is in the realm of international relations that differences and conflict is found, and the concept of Geopolitics has therefore migrated towards this arena—especially in its popular usage. Huntington's geopolitical model, especially the structures for North Africa and Eurasia, is largely derived from the "Intermediate Region" geopolitical model first formulated by Dimitri Kitsikis and published in 1978. Following Huntington and Mackinder, Zbigniew Brzezinski, in his book *The Grand Chessboard. American Primary and its Geostrategic Imperatives* (1997), developed a strategy for the US to retain its global hegemony. He renamed the Eurasian Heartland a Chessboard, and defined five countries as "pivots" to control the Eurasian landmass: France, Germany, Russia, China, and India. As a National Security Advisor for various presidents, Brzezinski's ideas have influenced US foreign policy since 1977.

German Geopolitics

German *Geopolitik* is characterized by the belief that life of States—being similar to those of human beings and animals—is shaped by scientific determinism and social Darwinism. German geopolitics develops the concept of *Lebensraum* (vital space) that is thought to be necessary to the development of a nation like a favourable natural environment would be for animals.

Friedrich Ratzel

Friedrich Ratzel (1844–1904), influenced by thinkers such as Darwin and zoologist Ernst Heinrich Haeckel, contributed to 'Geopolitik' by the expansion on the biological conception of geography, without a static conception of borders. Positing that states are organic and growing, with borders representing only a temporary stop in their movement, he held that the expanse of a state's borders is a reflection of the health of the nation—meaning that static countries are in decline. Ratzel published several papers, among which was the essay "Lebensraum" (1901) concerning biogeography. Ratzel created a foundation for the German variant of geopolitics, *geopolitik*. Influenced by the American geostrategist Alfred Thayer Mahan, Ratzel wrote of aspirations for German naval reach, agreeing that sea power was self-sustaining, as the profit from trade would pay for the merchant marine, unlike land power.

The geopolitical theory of Ratzel has been criticized as being too sweeping, and his interpretation of human history and geography being too simple and mechanistic. In his analysis of the importance of mobility, and the move from sea to rail transport, he failed to predict the revolutionary impact of air power. Critically, he also underestimated the importance of social organisation in the development of power.

The Association of German Geopolitik with Nazism

After World War I, the thoughts of Rudolf Kjellén and Ratzel were picked up and extended by a number of German authors such as Karl Haushofer (1869–1946), Erich Obst, Hermann Lautensach and Otto Maull. In 1923, Karl Haushofer founded the *Zeitschrift für Geopolitik* (Journal for Geopolitics), which was later used in the propaganda of Nazi Germany. The key concepts of Haushofer's Geopolitik were Lebensraum, autarky, pan-regions, and organic borders. States have, Haushofer argued, an undeniable right to seek natural borders which would guarantee autarky.

Haushofer's influence within the Nazi Party has recently been challenged, given that Haushofer failed to incorporate the Nazis' racial ideology into his work. Popular views of the role of geopolitics in the Nazi Third Reich suggest a fundamental significance on the part of the geo-politicians in the ideological orientation of the Nazi state. Bassin (1987) reveals that these popular views are in important ways misleading and incorrect. Despite the numerous similarities and affinities between the two doctrines, geopolitics was always held suspect by the National Socialist ideologists. This suspicion was understandable, for the underlying philosophical orientation of geopolitics did not comply with that of National Socialism. Geopolitics shared Ratzel's scientific materialism and determinism, and held that human society was determined by external influences—in the face of which qualities held innately by individuals or groups were of reduced or no significance. National Socialism rejected in principle both materialism and determinism and also elevated innate human qualities, in the form of a hypothesized 'racial character,' to the factor of greatest significance in the constitution of human society. These differences led after 1933 to friction and ultimately to open denunciation of geopolitics by Nazi ideologues. Nevertheless, German Geopolitik was discredited by its (mis)use in Nazi expansionist policy of World War II and has never achieved standing comparable to the pre-war period.

French Approach on Geopolitics

French geopolitical doctrines lie broadly in opposition to German Geopolitik and reject the idea of a fixed geography. French geography is focused on the evolution of polymorphic territories being the result of mankind actions. It also relies in the consideration of long time periods through refusal of taking specific events into account. This method has been theorized by Professor Lacoste according to three principles: Representation; Diachronie; and Diatopie.

In *The Spirit of the Laws*, Montesquieu outlined the view that man and societies are influenced by climate. He believed that hotter climates create hot-tempered people and colder climates aloof people, whereas the mild climate of France is ideal for political systems. Considered as one of the founders of French geopolitics, Élisée Reclus, is the author of a book considered as a reference in modern geography (Nouvelle Géographie universelle). Alike Ratzel, he considers geography through a global vision. However, in complete opposition to Ratzel's vision, Reclus considers geography not to be unchanging; it is supposed to evolve commensurately to the development of human society. His marginal political views resulted in his rejection by academia.

French geographer and geopolitician Jacques Ancel is considered to be the first theoretician of geopolitics in France, and gave a notable series of lectures at the Carnegie foundation and published "Géopolitique" in 1936. Like Reclus, Ancel rejects German determinist views on geopolitics (including Haushofer's doctrines).

Braudel's broad view used insights from other social sciences, employed the concept of the *longue durée*, and downplayed the importance of specific events. This method was inspired by the French geographer Paul Vidal de la Blache (who in turn was influenced by German thought, particularly that of Friedrich Ratzel whom he had met in Germany). Braudel's method was to analyse the interdependence between individuals and their environment. Vidalian geopolitics is based on varied forms of cartography and on *possibilism* (founded on a societal approach of geography—i.e. on the principle of spaces polymorphic faces depending from many factors among them mankind, culture, and ideas) as opposed to determinism.

Due to the influence of German *Geopolitik* on French geopolitics, the latter were for a long time banished from academic works. In the mid-1970s, Yves Lacoste—a French geographer who was directly inspired by Ancel, Braudel and Vidal de la Blache—wrote *La géographie, ça sert d'abord à faire la guerre* (Geography first use is war) in 1976. This book—which is very famous in France—symbolizes the birth of this new school of geopolitics (if not so far the first French school of geopolitics as Ancel was very isolated in the 1930s–40s). Initially linked with communist party evolved to a less liberal approach. At the end of the 1980s he founded the Institut Français de Géopolitique (French Institute for Geopolitics) that publishes the Hérodote revue. While rejecting the generalizations and broad abstractions employed by the German and Anglo-American traditions (and the *new geographers*), this school does focus on spatial dimension of geopolitics affairs on different levels of analysis. This approach emphazises the importance of multi-level (or multi-scales) analysis and maps at the opposite of critical geopolitics which avoid such tools. Lacoste proposed that every conflict (both local or global) can be considered from a perspective grounded in three assumptions:

1. Representation: Each group or individuals is the product of an education and is characterized by specific representations of the world or others groups or individuals. Thus, basic societal beliefs are grounded in their ethnicity or specific location. The study of representation is a common point with the more contemporary critical geopolitics. Both are connected with the work of Henri Lefebvre (La production de l'espace, first published in 1974)
2. Diachronie. Conducting an historical analysis confronting "long periods" and short periods as the prominent French historian Fernand Braudel suggested.
3. Diatopie: Conducting a cartographic survey through a multiscale mapping.

Connected with this stream, and former member of Hérodote editorial board, the French geographer Michel Foucher developed a long term analysis of international borders. He coined various neologism among them: *Horogenesis*: Neologism that describes the concept of studying the *birth of borders*, *Dyade*: border shared by two neighbouring states (for instance US territory has two terrestrial dyades : one with Canada and one with Mexico). The main book of this searcher "Fronts et frontières" (Fronts and borders) first published in 1991, without equivalent remains unfortunately untranslated in English. Michel Foucher is an expert of the African Union for borders affairs.

More or less connected with this school, Stéphane Rosière can be quoted as the editor in Chief of the online journal *L'Espace politique*, this journal created in 2007 became the most prominent French journal of political geography and Geopolitics with Hérodote.

A much more conservative stream is personified by François Thual was a French expert in geopolitics, and a former official of the Ministry of Civil Defence. Thual taught geopolitics of the religions at the French War College, and has written thirty books devoted mainly to geopolitical method and its application to various parts of the world. He is particularly interested in the Orthodox, Shiite, and Buddhist religions, and in troubled regions like the Caucasus. Connected with F. Thual, Aymeric Chauprade, former professor of geopolitics at the French War College and now member of the extreme-right party "Front national", subscribes to a supposed "new" French school of geopolitics which advocates above all a return to *realpolitik* and "clash of civilization" (Huntington). The thought of this school is expressed through the French Review of Geopolitics (headed by Chauprade) and the International Academy of Geopolitics. Chauprade is a supporter of a Europe of nations, he advocates a European Union excluding Turkey, and a policy of compromise with Russia (in the frame of a Eurasian alliance which is *en vogue* among European extreme-right politists) and supports the idea of a multipolar world—including a balanced relationship between China and the U.S.

Russian Geopolitics

The modern day Russian geopolitics is centred on Eurasianist tradition and is highly interlinked with politics. The trauma of the disintegration of the Soviet Union left behind various views ranging from moderate – stressing the unique position of Russia between Europe and Asia – to more extreme – arguing for Greater Russia aspirations (renaissance of Russian empire in the borders of the former Soviet Union), in line with the expansionist views of Alexander Dugin.

Meta-Geopolitics

The framework of *Meta*-geopolitics, proposed by Nayef Al-Rodhan combines traditional and new dimensions of geopolitics to offer a multidimensional view of power and power relationships. In this framework, the importance of geography is superseded by the combination of hard- and soft- power tools that states can employ to preserve and obtain power. *Meta*-geopolitics defines seven key dimensions of state power that include social and health issues, domestic politics, economics, environment, science and human potential, military and security issues, and international diplomacy. The *Meta*-geopolitics framework allows for the assessment of relative strengths and weaknesses as well as predictions about future trends. Furthermore, while this analytical grid is relevant for states, it also applies to private and transnational entities, which are playing an increasingly important role in contemporary geopolitics.

Non-Interventionism

Non-interventionism, the geopolitical policy whereby a nation seeks to avoid alliances with other nations in order to avoid being drawn into wars not related to direct territorial self-defence. An original more formal definition is that Non-intervention is a policy characterized by the absence of interference by a state or states in the external affairs of another state without its consent, or in its internal affairs with or without its consent. This is based on the grounds that a state should not interfere in the internal politics of another state, based upon the principles of state sovereignty and self-determination. A similar phrase is “strategic independence”. Historical examples of supporters of non-interventionism are US Presidents George Washington and Thomas Jefferson, who both favoured nonintervention in European Wars while maintaining free trade. Other proponents include United States Senator Robert Taft and United States Congressman Ron Paul.

Democracy and Geopolitics

State geopolitical interests can be also seen as anti-democratic in nature, as they lie very away from the interests of the general public. Vietnam War is a very prominent example of this: in by 1970 only a third of Americans believed that the U.S. had not made a mistake by sending troops to fight in Vietnam, however USA continued to defend its geopolitical interests by military intervention in Vietnam for another 5 years, till 1975. Another example is the presence of US troops in Afghanistan, which in 2013 was supported by only 20% of US population, but US military presence still continues till 2014.

Space vs. Place

In the humanistic geography space and place are important concepts. Concepts that in this approach doesn’t mean the same. Space is something abstract, without any substantial meaning. While place refers to how people are aware of/attracted to a certain piece of space. A place can be seen as space that has a meaning. The underlying theory for this way of thinking is the phenomenology, which tries to find the essential features of experiences in the direct and indirect experiences.

Space and Place According to Yi-Fu Tuan and Relph

Two important thinkers who give a contribution in defining space, place and the differences between these concepts are philosopher Yi-Fu Tuan and geographer Edward Relph. They have almost the same ideas about the distinction between space an place.

Yi-Fu Tuan searched for the meaning of place, space and environment. According to Tuan the difference between 'space' and 'place can be described in the extent to which human beings have given meaning to a specific area. Meaning can be given or derived from in area in two different ways, namely:

- In an direct and intimate way, for example through the senses such as vision, smell, sense and hearing.
- An in an indirect and conceptual way mediated by symbols, arts etc. (Tuan, 1977, p. 6).

'Space' can be described as a location which has no social connections for a human being. No value has been added to this space. According to Tuan (1977, p.164-165) it is an open space, but may marked off and defended against intruders (Tuan, 1977, p. 4). It does not invite or encourage people to fill the space by being creative. No meaning has been described to it. It is more or less abstract (Tuan, 1977, p. 6).

'Place' is in contrary more than just a location and can be described as a location created by human experiences. The size of this location does not matter and is unlimited. It can be a city, neighbourhood, a region or even a classroom et cetera. In fact 'place' exists of 'space' that is filled with meanings and objectives by human experiences in this particular space. Places are centres where people can satisfy there biological needs such as food, water etc. (Tuan, 1977, p. 4). According to Tuan (1977, p. 6) a 'place' does not exist of observable boundaries and is besides a visible expression of a specific time period. Examples are arts, monuments and architecture.

Tuan was convinced that people give or derive meaning from the world's geography and organise the world around themselves (Cloke, Philo & Sadler, 1991, p. 76-77). This also explains that the meaning we give to 'space' correlates with the distance from the human to this 'place' (Cloke , P., Philo, et. al., 1991, p. 79).As Tuan (1977, p. 3) says: *"space is freedom, place is security"*.

Edward Relph broadly has the same ideas as Tuan. An addition that can be made is that Relph describe the relationship between people and their places as detached and at distance, through which people are able to reflect on this relationship (Cloke, Philo, et al., 1992). In his work Relph tries to maintain the relation between space and place and not present them as separated concepts. Because the quality of place is that it has the power to order and to focus human intentions, experiences and actions spatially (Seamon & Sowers, 2008). *"So space and place are dialectically structured in human environmental experience, since our understanding of space is related to the places we inhabit, which in turn derive meaning from their spatial context"* (Seamon & Sowers, 2008, p.44).

Social Exclusion

Social exclusion (or marginalization) is social disadvantage and relegation to the fringe of society. It is a term used widely in Europe, and was first used in France. It is used across disciplines including education, sociology, psychology, politics and economics.

Social exclusion is the process in which individuals or entire communities of people are systematically blocked from (or denied full access to) various rights, opportunities and resources that are normally available to members of a different group, and which are fundamental to social integration within that particular group (e.g., housing, employment, healthcare, civic engagement, democratic participation, and due process).

Alienation or disenfranchisement resulting from social exclusion is often connected to a person's social class, educational status, childhood relationships, living standards, or personal choices in fashion.

Such exclusionary forms of discrimination may also apply to people with a disability, minorities, members of the LGBT community, "seniors", or young people. Anyone who appears to deviate in any way from the "perceived norm" of a population may thereby become subject to coarse or subtle forms of social exclusion.

The outcome of social exclusion is that affected individuals or communities are prevented from participating fully in the economic, social, and political life of the society in which they live.

Most of the characteristics listed in this article are present together in studies of social exclusion, due to exclusion's multidimensionality.

Another way of articulating the definition of social exclusion is as follows:

Social exclusion is a multidimensional process of progressive social rupture, detaching groups and individuals from social relations and institutions and preventing them from full participation in the normal, normatively prescribed activities of the society in which they live.

One model to conceptualize social exclusion and inclusion is that they are on a continuum on a vertical plane below and above the 'social horizon'. According to this model, there are ten social structures that impact exclusion and can fluctuate over time: race, geographic location, class structure, globalization, social issues, personal habits and appearance, education, religion, economics and politics.

In an alternative conceptualization, social exclusion theoretically emerges at the individual or group level on four correlated dimensions: insufficient access to social rights, material deprivation, limited social participation and a lack of normative integration. It is then regarded as the combined result of personal risk factors (age, gender, race); macro-societal changes (demographic, economic and labour market developments, technological innovation, the evolution of social norms); government legislation and social policy; and the actual behaviour of businesses, administrative organisations and fellow citizens.

An inherent problem with the term, however, is the tendency of its use by practitioners who define it to fit their argument.

Individual Exclusion

"The marginal man...is one whom fate has condemned to live in two societies and in two, not merely different but antagonistic cultures....his mind is the crucible in which two different and refractory cultures may be said to melt and, either wholly or in part, fuse."

Social exclusion at the individual level results in an individual's exclusion from meaningful participation in society. An example is the exclusion of single mothers from the welfare system prior to welfare reforms of the 1900s. The modern welfare system is based on the concept of entitlement to the basic means of being a productive member of society both as an organic function of society and as compensation for the socially useful labour provided. A single mother's contribution to society is not based on formal employment, but on the notion that provision of welfare for children is a necessary social expense. In some career contexts, caring work is devalued and motherhood is seen as a barrier to employment. Single mothers were previously marginalized in spite of their significant role in the socializing of children due to views that an individual can only contribute meaningfully to society through "gainful" employment as well as a cultural bias against unwed mothers. Today the marginalization is primarily a function of class condition.

More broadly, many women face social exclusion. Moosa-Mitha discusses the Western feminist movement as a direct reaction to the marginalization of white women in society. Women were excluded from the labour force and their work in the home was not valued. Feminists argued that men and women should equally participate in the labour force, in the public and private sector, and in the home. They also focused on labour laws to increase access to employment as well as to recognise child-rearing as a valuable form of labour. Today, women are still marginalized from executive positions and continue to earn less than men in upper management positions.

Another example of individual marginalization is the exclusion of individuals with disabilities from the labour force. Grandz discusses an employer's viewpoint about hiring individuals living with disabilities as jeopardizing productivity, increasing the rate of absenteeism, and creating more accidents in the workplace. Cantor also discusses employer concern about the excessively high cost of accommodating people with disabilities. The marginalization of individuals with disabilities is prevalent today, despite the legislation intended to prevent it in most western countries, and the academic achievements, skills and training of many disabled people.

There are also exclusions of lesbian-gay-bisexual-transgender (LGBT) and other intersexual people because of their sexual orientations and gender identities. The Yogyakarta Principles require that the states and communities abolish any stereotypes about LGBT people as well as stereotyped gender roles.

"Isolation is common to almost every vocational, religious or cultural group of a large city. Each develops its own sentiments, attitudes, codes, even its own words, which are at best only partially intelligible to others."

Community Exclusion

Many communities experience social exclusion, such as racial (e.g., black) and economic (e.g., Romani) communities.

One example is the Aboriginal community in Australia. Marginalization of Aboriginal communities is a product of colonization. As a result of colonialism, Aboriginal communities lost their land, were forced into destitute areas, lost their sources of livelihood, and were excluded from the labour market. Additionally, Aboriginal communities lost their culture and values through forced assimilation and lost their rights in society. Today various Aboriginal communities continue to be marginalized from society due to the development of practices, policies and programs that "met the needs of white people and not the needs of the marginalized groups themselves". Yee also connects marginalization to minority communities, when describing the concept of whiteness as maintaining and enforcing dominant norms and discourse.

Professional Exclusion

Some intellectuals and thinkers are marginalised because of their dissenting, radical or controversial views on a range of topics, including HIV/AIDS, climate change, evolution, alternative medicine, green energy, or third world politics. Though fashionable for a time to some, they are more widely regarded as intellectual freethinkers and dissidents whose ideas and views run against those of the mainstream. At times they are marginalised and abused, often systematically ostracized by colleagues, and in some cases their work ridiculed or banned from publication. Examples include Immanuel Velikovsky, Peter Duesberg, Susan George, Martin Fleischman, Stanley Pons, Fred Hoyle, James Lovelock, E. F. Schumacher.

Other Contributors

Social exclusion has many contributors. Major contributors include race, income, employment status, social class, geographic location, personal habits and appearance, education, religion and political affiliation.

Global and Structural

Globalization (global-capitalism), immigration, social welfare and policy are broader social structures that have the potential to contribute negatively to one's access to resources and services, resulting in the social exclusion of individuals and groups. Similarly, increasing use of information technology and company outsourcing have contributed to job insecurity and a widening gap between the rich and the poor. Alphonse, George & Moffat (2007) discuss how globalization sets forth a decrease in the role of the state with an increase in support from various "corporate sectors resulting in gross inequalities, injustices and marginalization of various vulnerable groups" (p. 1). Companies are outsourcing, jobs are lost, the cost of living continues to rise, and land is being expropriated by large companies. Material goods are made in large abundances and sold at cheaper costs, while in India for example, the poverty line is lowered in order to mask the number of individuals who are actually living in poverty as a result of globalization.

Globalization and structural forces aggravate poverty and continue to push individuals to the margins of society, while governments and large corporations do not address the issues (George, P, SK8101, lecture, October 9, 2007).

Certain language and the meaning attached to language can cause universalizing discourses that are influenced by the Western world, which is what Sewpaul (2006) describes as the "potential to dilute or even annihilate local cultures and traditions and to deny context specific realities" (p. 421). What Sewpaul (2006) is implying is that the effect of dominant global discourses can cause individual and cultural displacement, as well as an experience of "de-localization", as individual notions of security and safety are jeopardized (p. 422). Insecurity and fear of an unknown future and instability can result in displacement, exclusion, and forced assimilation into the dominant group. For many, it further pushes them to the margins of society or enlists new members to the outskirts because of global-capitalism and dominant discourses (Sewpaul, 2006).

With the prevailing notion of globalization, we now see the rise of immigration as the world gets smaller and smaller with millions of individuals relocating each year. This is not without hardship and struggle of what a newcomer thought was going to be a new life with new opportunities. Ferguson, Lavalette, & Whitmore (2005) discuss how immigration has had a strong link to access of welfare support programs. Newcomers are constantly bombarded with the inability to access a country's resources because they are seen as "undeserving foreigners" (p. 132). With this comes a denial of access to public housing, health care benefits, employment support services, and social security benefits (Ferguson et al., 2005). Newcomers are seen as undeserving, or that they must prove their entitlement in order to gain access to basic support necessities. It is clear that individuals are exploited and marginalized within the country they have emigrated (Ferguson et al., 2005).

Welfare states and social policies can also exclude individuals from basic necessities and support programs. Welfare payments were proposed to assist individuals in accessing a small amount of material wealth (Young, 2000). Young (2000) further discusses how "the provision of the welfare itself produces new injustice by depriving those dependent on it of rights and freedoms that others have...marginalization is unjust because it blocks the opportunity to exercise capacities in socially defined and recognised way" (p. 41). There is the notion that by providing a minimal amount of welfare support, an individual will be free from marginalization. In fact, welfare support programs further lead to injustices by restricting certain behaviour, as well the individual is mandated to other agencies. The individual is forced into a new system of rules while facing social stigma and stereotypes from the dominant group in society, further marginalizing and excluding individuals (Young, 2000). Thus, social policy and welfare provisions reflect the dominant notions in society by constructing and reinforcing categories of people and their needs. It ignores the unique-subjective human essence, further continuing the cycle of dominance (Wilson & Beresford, 2000).

Unemployment

Whilst recognising the multi-dimensionality of exclusion, policy work undertaken at European Union level focuses on unemployment as a key cause of, or at least correlating with, social exclusion. This is because in modern societies, paid work is not only the principal source of income with which to buy services, but is also the fount of individuals' identity and feeling of self-worth. Most people's social networks and sense of embeddedness in society also revolve around their work. Many of the indicators of extreme social exclusion, such as poverty and homelessness, depend on monetary income which is normally derived from work. Social exclusion can be a possible result of long-term unemployment, especially in countries with weak welfare safety nets. Much policy to reduce exclusion thus focuses on the labour market:

- On the one hand, to make individuals at risk of exclusion more attractive to employers, i.e. more "employable".
- On the other hand, to encourage (and/or oblige) employers to be more inclusive in their employment policies.

The EU's EQUAL Community Initiative investigated ways to increase the inclusiveness of the labour market. Work on social exclusion more broadly is carried out through the Open Method of Coordination (OMC) among the Member State governments.

Transportation

In some circumstances, transport may be a factor in social exclusion - for instance, if lack of access to public transport or a vehicle prevents a person from getting to a job, training course, job centre or doctor's surgery, entertainment venues. Some schemes therefore promote accessibility, for instance:

- By ensuring public transport is available.
- By subsidising the purchase of a scooter, which is relevant to young people living in rural areas, or a car, or a bicycle, etc.

Religion

Individuals and communities can be socially excluded on the basis of their religious beliefs. Significant historical examples include the Spanish inquisition, Jewish life under Nazi rule, Christians in Egypt and in other Muslim-majority countries, and Islamophobia in the United States after the September 11 attacks.

Within some religious traditions, individuals who are seen as deviating from a religious teaching may be excluded. One example is excommunication in Christianity. Excommunication is generally justified by the religion's internal code, such as Douglas A. Jacoby's interpretation of 1 Corinthians 5:11 and Titus 3:9-11 as evidence that members can be excluded from fellowship for matters perceived within the church as grave sin without a religiously acceptable repentance.

Links between Exclusion and Other Issues

The problem of social exclusion is usually tied to that of equal opportunity, as some people are more subject to such exclusion than others. Marginalisation of certain groups is a problem even in many economically more developed countries, including the United Kingdom and the United States, where the majority of the population enjoys considerable economic and social opportunities.

Since social exclusion may lead to one being deprived of one's citizenship, some authors (Philippe Van Parijs, Jean-Marc Ferry, Alain Caillé, André Gorz and Axel Wolz) have proposed a basic income, which would impede exclusion from citizenship. The concept of a Universal Unconditional Income, or social salary, has been disseminated notably by the Green movement in Germany.

In the last few years, there has been research focused on possible connections between exclusion and brain function. Studies published by the University of Georgia and San Diego State University found that exclusion can lead to diminished brain functioning and poor decision making. Such studies corroborate with earlier beliefs of sociologists. The effect of exclusion may likely correlate with such things as substance abuse and crime.

Social Inclusion

Social inclusion, the converse of social exclusion, is affirmative action to change the circumstances and habits that lead to (or have led to) social exclusion. The World Bank defines social inclusion as the process of improving the ability, opportunity, and dignity of people, disadvantaged on the basis of their identity, to take part in society.

Social Inclusion ministers have been appointed, and special units established, in a number of jurisdiction around the world. The first Minister for Social Inclusion was Premier of South Australia Mike Rann, who took the portfolio in 2004. Based on the UK's Social Exclusion Unit, established by Prime Minister Tony Blair in 1997, Rann established the Social Inclusion Initiative in 2002. It was headed by Monsignor David Cappo and was serviced by a unit within the department of Premier and Cabinet. Cappo sat on the Executive Committee of the South Australian Cabinet and was later appointed Social Inclusion Commissioner with wide powers to address social disadvantage. Cappo was allowed to roam across agencies given that most social disadvantage has multiple causes necessitating a "joined up" rather than a single agency response. The Initiative drove a big investment by the South Australian Government in strategies to combat homelessness, including establishing Common Ground, building high quality inner city apartments for "rough sleeping" homeless people, the Street to Home initiative and the ICAN flexible learning program designed to improve school retention rates. It also included major funding to revamp mental health services following Cappo's "Stepping Up" report, which focused on the need for community and intermediate levels of care and an overhaul of disability services. In 2007 Australian Prime Minister Kevin Rudd appointed Julia Gillard as the nation's first Social Inclusion Minister.

Consequences of Social Exclusion

Crime: Sociologists see strong links between crime and social exclusion in industrialized societies such as the United States. Growing crime rates may reflect the fact that a growing number of people do not feel valued in the societies in which they live. The socially excluded population are in favour of illegal means of fulfilling their goals and motives in life as they have no other way to fit into a society that will not accept them. Crime is favoured over the political system or community organisation. Young people increasingly grow up without guidance and support from the adult population. Young people also face diminishing job opportunities to sustain a livelihood. This can cause a sense of willingness to turn to illegitimate means of sustaining a desired lifestyle.

Health

In gay men, results of psychoemotional damage from marginalization from both heterosexual society and from within mainstream homosexual society include bug chasing (purposeful acts to acquire HIV), suicide, and drug addiction.

In Philosophy

The marginal, the processes of marginalisation, etc. bring specific interest in postmodern and postcolonial philosophy and social studies. Postmodernism question the "centre" about its authenticity and postmodern sociology and cultural studies research marginal cultures, behaviours, societies, the situation of the marginalized individual, etc.

Linda Hutcheon describes postmodernism itself as "intertextual, parodic, contradictory, provisional, heterogeneous, transgressive of generic divisions, ex-centric and *marginal*".

Implications for Social Work Practice

Upon defining and describing marginalization as well as the various levels in which it exists, one must now explore its implications for social work practice. Mullaly (2007) describes how "the personal is political" and the need for recognising that social problems are indeed connected with larger structures in society, causing various forms of oppression amongst individuals resulting in marginalization (p. 262). It is also important for the social worker to recognise the intersecting nature of oppression. A non-judgmental and unbiased attitude is necessary on the part of the social worker. The worker must begin to understand oppression and marginalization as a systemic problem, not the fault of the individual (Mullaly, 2007).

Working under an Anti-oppression perspective would then allow the social worker to understand the lived, subjective experiences of the individual, as well as their cultural, historical and social background. The worker should recognise the individual as political in the process of becoming a valuable member of society and the structural factors that contribute to oppression and marginalization (Mullaly, 2007). Social workers must take a firm stance on naming and labelling global forces that impact individuals and communities who are then left with no support, leading to marginalization or further marginalization from the society they once knew (George, P, SK8101, lecture, October 9, 2007).

The social worker should be constantly reflexive, work to raise the consciousness, empower, and understand the lived subjective realities of individuals living in a fast-paced world, where fear and insecurity constantly subjugate the individual from the collective whole, perpetuating the dominant forces, while silencing the oppressed (Sakamoto and Pitner, 2005).

Some individuals and groups who are not professional social workers build relationships with marginalized persons by providing relational care and support, for example, through homeless ministry. These relationships validate the individuals who are marginalized and provide them a meaningful contact with the mainstream.

Juridical Concept

There are countries, Italy for example, that have a legal concept of *social exclusion*. In Italy, "*esclusione sociale*" is defined as poverty combined with social alienation, by the statute n. 328 (11-8-2000), that instituted a state investigation commission named "*Commissione di indagine sull 'Esclusione Sociale*" (CIES) to make an annual report to the government on legally expected issues of social exclusion.

The Vienna Declaration and Programme of Action, a document on international human rights instruments affirms that "extreme poverty and social exclusion constitute a violation of human dignity and that urgent steps are necessary to achieve better knowledge of extreme poverty and its causes, including those related to the program of development, in order to promote the human rights of the poorest, and to put an end to extreme poverty and social exclusion and promote the enjoyment of the fruits of social progress. It is essential for States to foster participation by the poorest people in the decision making process by the community in which they live, the promotion of human rights and efforts to combat extreme poverty."

Geography and Wealth

Geography and wealth have long been perceived as correlated attributes of nations. The continents along the equator, Africa and India, are the poorest. Even within Africa and India this effect can be seen, as the nations farthest from the equator are wealthier. In Africa the wealthiest nations are the three on the southern tip of the continent, South Africa, Botswana, and Namibia, and the countries of North Africa. Similarly in Latin America Argentina, Southern Brazil, Chile, and Uruguay have long been the wealthiest. Within Asia, Indonesia, located on the equator, is among the poorest. Within Central Asia, Kazakhstan is wealthier than other former Soviet Republics which border it to the south, like Uzbekistan. The wealthiest nations of the world with the highest standard of living tend to be those at the northern extreme of areas open to human habitation, Canada, and the Nordic Countries. Within the wealthy continents, and even within large countries, wealth increases with distance from the equator. Southern Europe has long been poorer than its northern counterpart, as has the Southern United States than the northeast counterpart.

Researchers at Harvard's Centre for International Development (CID) found in 2001 that only three tropical economies — Hong Kong, Singapore, and part of Taiwan — were classified as high-income by the World Bank, while all countries within regions zoned as temperate had either middle- or high-income economies.

There are exceptions, for example, Russia is less well off than the United States, or even Australia; the latter country is wealthier than Southern Europe. Also, within Russia, Moscow and especially St. Petersburg are wealthier than, for example, Siberia. Similarly, Germany's poorest regions are situated in the North East. The wealthiest nations of Central America are generally those closest to the equator, namely Panama and Costa Rica, and Mongolia is poorer than China. Also, Alaska and Canada are poorer than the mainland United States. In Canada, the poorest region is in the North. In China, the poorest regions are in the West and the North East. And in south-east Asia, the richest nations are Malaysia, Thailand and Singapore, straddling the equator.

Measurement

Most of the recent studies use national GDP per capita, as measured by the World Bank and the International Monetary Fund, as the unit of comparison. Intra-national comparisons use their own data, and political divisions such as states or provinces then delineate the study areas.

Explanations

Historic: One of the first to describe and assess the phenomenon was the French philosopher Montesquieu, who asserted in the 18th century that "cold air constringes (*sic*) the extremities of the external fibres of the body; this increases their elasticity, and favours the return of the blood from the extreme parts to the heart. It contracts those very fibres; consequently it increases also their force. On the contrary, warm air relaxes and lengthens the extremes of the fibres; of course it diminishes their force and elasticity. People are therefore more vigorous in cold climates."

The 19th century historian Henry Thomas Buckle wrote that "climate, soil, food, and the aspects of nature are the primary causes of intellectual progress,—the first three indirectly, through determining the accumulation and distribution of wealth, and the last by directly influencing the accumulation and distribution of thought, the imagination being stimulated and the understanding subdued when the phenomena of the external world are sublime and terrible, the understanding being emboldened and the imagination curbed when they are small and feeble."

Cultural Innovation

Physiologist Jared Diamond was inspired to write his Pulitzer Prize-winning work Guns, Germs, and Steel by a question posed by a New Guinean politician: why were Europeans so much wealthier than his people? In this book, Diamond argues that the Europe-Asia land mass is particularly favourable for the transition of societies from hunter-gatherer to farming communities. This continent stretches much further along the same lines of latitude than any of the other continents. Since it is much easier to transfer a domesticated species along the same latitude than it is to move it to a warmer or colder climate, any species developed at a particular latitude will be transferred across the continent in a relatively short amount of time. Thus the inhabitants of the Eurasian continent have had a built-in advantage in terms of earlier development of farming, and a greater range of plants and animals from which to choose.

Diamond notes that modern technologies and institutions were designed primarily in a small area of northwestern Europe. After the Scientific Revolution in Europe in the 16th century, the quality of life increased and wealth began to spread to the middle class. This included agricultural techniques, machines, and medicines. These technologies and models readily spread to areas colonized by Europe which happened to be of similar climate, such as North America and Australia. As these areas also became centres of innovation, this bias was further enhanced. Technologies from automobiles to power lines are more often designed for colder and drier regions, since most of their customers are from these regions.

The book goes on to document a feedback effect of technologies being designed for the wealthy, which makes them more wealthy and thus more able to fund technological development. He notes that the far north has not always been the wealthiest latitude; until only a few centuries ago, the wealthiest belt stretched from Southern Europe through the Middle East, northern India and southern China. A dramatic shift in technologies, beginning with ocean-going ships and culminating in the Industrial Revolution, saw the most developed belt move north into northern Europe, China, and the Americas. Northern Russia became a superpower while southern India became impoverished and colonized. Diamond argues that such dramatic changes demonstrate that the current distribution of wealth is not due to immutable factors such as climate or race, citing the early emergence of agriculture in ancient Mesopotamia as evidence.

Diamond also notes the feedback effect in the growth of medical technology, stating that more research money goes into curing the ailments of northerners.

Disease

Ticks, mosquitos, and rats, which continue to be important disease vectors, benefit from warmer temperatures and higher humidities. There has long been a malarial belt spanning the equatorial portions of the globe; the disease is especially deadly to children under the age of five. Notably it has been almost impossible for most forms of northern livestock to thrive due to the endemic presence of the tsetse fly.

Jared Diamond has linked domestication of animals in Europe and Asia to the development of diseases that enabled these countries to conquer the inhabitants of other continents. The close association of people in Eurasia with their domesticated animals provided a vector for the rapid transmission of diseases. Inhabitants of lands with few domesticated species were never exposed to the same range of diseases, and so, at least on the American continents, succumbed to diseases introduced from Eurasia. These effects were exhaustively discussed in William McNeill's book *Plagues and Peoples*.

The 2001 Harvard study mentions high infant mortality as another factor; since birth rates usually increase in compensation, women may delay their entry into the workforce to care for their younger children. The education of the surviving children then becomes difficult, perpetuating a cycle of poverty.

Other

- In "Climate and Scale in Economic Growth," William A. Masters and Margaret S. McMillan of Purdue University and Tufts University hypothesize that the disparity is partially due to the effects of frost in increasing soil fertility.
- Hernando Zuleta, of the Universidad del Rosario, has proposed that where output fluctuations are more profound, i.e. areas that experience winter, saving is more pronounced, which leads to the adoption or creation of capital-intensive technologies.
- Daron Acemoglu, Simon Johnson, and James A. Robinson of MIT argued in 2001 that in places where Europeans faced high mortality rates, they could not settle and were more likely to

set up exploitative institutions. These institutions offered no protection for private property or checks and balances against government expropriation. They assert that after controlling for the effect of institutions, countries in Africa or those closer to the equator do not have lower incomes. This work has been disputed by David Albouy, who argues that European mortality rates in the study were badly mismeasured, falsely supporting its conclusion.

- The authors of "Brain size, cranial morphology, climate, and time machines" assert that colder climates increase brain size, resulting in an intelligence differential.

Impact of Global Warming on Wealth

In a 2006 paper discussing the potential impact of global warming on wealth, John K. Horowitz of the University of Maryland predicted that a 2-degree Fahrenheit temperature increase across all countries would cause a decrease of 2 to 6 percent in world GDP, with a best estimate of around 3.5 percent. United Nations Secretary-General Ban Ki-moon has expressed concern that global warming will exacerbate the existing poverty in Africa.

Bibliography

Ashraf, A.: *Political Sociology : A New Grammar of Politics*, Universities Press, Delhi, 1983.

Balshaw, Maria, and Liam Kennedy: *Urban Space and Representation.* Sterling, VA: Pluto Press, 2000.

Barnes, Trevor J., and James S. Duncan: *Writing Worlds: Discourse, Text, and Metaphor in the Representation of Landscape.* London and New York: Routledge, 1992.

Belden C.: *Landscapes of the Sacred: Geography and Narrative in American Spirituality*, New York: Paulist Press, 1988.

Buttimer, Anne: *International Encyclopedia of the Social Sciences*, MacMillan, New York, 1968.

Carrillo, J., Lung, Y., and van Tulder, R. : *Cars, Carriers of Regionalism*, Palgrave Macmillan, New York, 2004.

Cronon, William: *Changes in the Land: Indians, Colonists, and the Ecology of New England.* New York: Hill and Wang, 1983.

Damon, William : *Social and Personality Development, Infancy through Adolescence,* New York: Norton, 1983.

Davis, Mike: *Ecology of Fear: Los Angeles and the Imagination of Disaster.* New York: Vintage, 1999.

Dibbern, Jens : *The Sourcing of Application Software Services, Empirical Evidence of Cultural, Industry and Functional Differences,* Physica Verlag Heidelberg, 2004.

Elangovan K. : *GIS, Fundamentals, Applications and Population Geography*, New India Publishing Agency, New Delhi 2006.

Fran Tonkiss: *Contemporary Economic Sociology: Globalization, Production, Inequality*, Manohar, Delhi, 2008.

Gregory, Derek and John Urry: *Social Relations and Spatial Structures*, MacMillan, Basingstoke, 1985.

Harvey, David: *Explanation in Geography.* London: Edward Arnold, 1969.

Howes, Elaine V. : *Connecting Girls and Science: Constructivism, Feminism, and Science Education Reform*, New York, Teachers College Press, 2002.

Jackson, Peter and Susan J. Smith: *Exploring Social Geography*, Allen & Unwin, Boston, London, 1984.

Jagmohan Diwan: *Fundamentals of Rural Sociology*, Cyber Tech Pub, Delhi, 2009.

James, T.: *The Best Poor Man's Country: A Geographical Study of Early Southeastern Pennsylvania*. Baltimore: Johns Hopkins Press, 1972.

Jeff Hass: *Economic Sociology: An Introduction*, Manohar, Delhi, 2008.

Karan Raj: *Sarup Dictionary of Sociology*, Sarup, Delhi, 2002.

Laura Kramer: *The Sociology of Gender : A Brief Introduction*, Rawat, Delhi, 2004.

Lewis, Peirce. *New Orleans: The Making of an Urban Landscape*. Cambridge, MA: Ballinger Publishing Co., 1976. Annotation

M Francis Abraham: *Contemporary Sociology : An Introduction to Concepts and Theories*, Oxford University Press, New York, 2006.

Macionis, John J., Vincent Parillo: *Urban Sociology*, Boston, Allyn and Bacon, 2001.

Masood Ali Khan: *Cultural Sociology of India*, Arise Pub, Delhi, 2006.

Michael P.: *The Making of the American Landscape*. Boston: Unwin Hyman, 1990.

Richard SCASE : *Managing Creativity*, Open University Press, 2000.

Ritchie, J.: *Landscape and Material Life in Franklin County, Massachusetts, 1770-1860*. Knoxville: University of Tennessee Press, 1991.

Stephen, D.: *Fields of Vision: Landscape Imagery and National Identity in England and the United States*. Cambridge: Polity, 1993.

Smith, Susan J.: *The Sage Handbook of Social Geographies*, Sage, London, 2010.

Victor, A. : *Geography of Art and Culture*, Contributions to Economic Analysis, Elsevier, 2004.

Valentine, Gill: *Social Geographies: Space and Society*, Prentice Hall, New York, 2001.

Watkins, Eric: *The Middle Eastern Environment*, London, The British Society for Middle Eastern Studie, 1995.

Werlen, Benno: *Society, Action and Space: An Alternative Human Geography*, Routledge, London, New York, 1993.

Index

❑❑❑